Do You See Ice?

Do You See Ice?

Inuit and Americans at Home and Away

KAREN ROUTLEDGE

The University of Chicago Press
Chicago and London

PUBLICATION OF THIS BOOK HAS BEEN AIDED BY A GRANT FROM THE
BEVINGTON FUND.

The University of Chicago Press, Chicago 60637
The University of Chicago Press, Ltd., London
© 2018 by The University of Chicago
All rights reserved. No part of this book may be used or reproduced in any manner whatsoever without written permission, except in the case of brief quotations in critical articles and reviews. For more information, contact the University of Chicago Press, 1427 E. 60th St., Chicago, IL 60637.
Published 2018
Printed in the United States of America

27 26 25 24 23 22 21 20 19 18   1 2 3 4 5

ISBN-13: 978-0-226-58013-5 (cloth)
ISBN-13: 978-0-226-58027-2 (e-book)
DOI: https://doi.org/10.7208/chicago/9780226580272.001.0001

Library of Congress Cataloging-in-Publication Data

Names: Routledge, Karen, author.
Title: Do you see ice? : Inuit and Americans at home and away / Karen Routledge.
Description: Chicago ; London : The University of Chicago Press, 2018. | Includes bibliographical references and index.
Identifiers: LCCN 2018001640 | ISBN 9780226580135 (cloth : alk. paper) | ISBN 9780226580272 (e-book)
Subjects: LCSH: Inuit—Nunavut—Cumberland Sound. | Americans—Nunavut—Cumberland Sound.
Classification: LCC E99.E7 R698 2018 | DDC 971.9/520049712—dc23
LC record available at https://lccn.loc.gov/2018001640

♾ This paper meets the requirements of ANSI/NISO Z39.48-1992 (Permanence of Paper).

Contents

Acknowledgments vii
Prologue: On the Ice xi

1	Americans in Cumberland Sound	1
2	Inuit in the United States	35
3	Americans and Inuit in the High Arctic	77
4	Inuit in Cumberland Sound	110

Epilogue: At Home 149
Appendix: Methodological Essay 157
Notes 161
Works Cited 197
Index 217

Acknowledgments

ᑕᐁᕗᙳᓕᕆᖅ ᖅᑯᔭᕐᒦᐱᐊᖑᓯᓂᐊᓕᖅᒪ ᐃᓇᑐᖃᓯᓂᖅ ᑐᓯᕐ ᐊᓅᔾ ᑎᐊᓗᔾ
ᐊᐱᖅᓱᖃᑦᑕᕐᓯᑎᓐᓇᓂᖅ ᐸᓇᓂᖅᑐᒥ, ᑕᐱᒃᑯᐊᖑᓕᐤ ᐊᒥᔪᑦ ᐃᓅᔪᓐᓃᖅᓯᒪᓕᕐᒥᒪᑕ.
ᖅᑯᔭᓇᒦᖅᑦᖅᑲ ᐃᓅᔭ ᓇᓯᓕᐤ, ᔭᒥᓯ ᒪᐃᑉ, ᐸᐅᓗᔾ ᐋᐊ, ᐳᔾ ᐋᐊ, ᐄᐊ ᐊᓂᓐᓂᓕᐊᖅ,
ᑕᐃᓯ ᑎᐊᓕ, ᓴᐅᓪᓗ ᓇᒃᓴᖅ, ᐱᑐᕈᓯ ᖃᓕᐤ ᐊᒻᓗ ᐃᓕᓴᐱ ᐃᓱᓗᑕᖅ. ᐅᑯᐊᑦᑲᐅᖅ
ᑲᑎᒪᐅᖅᒥᓯᒥᓂᕈᓕᖅᓱᓚᐊᖅᑲ, ᐃᑦᑐᐊᖑᓕᑦ ᐊᖅᓴᖅ, ᐸᐅᓗᔾ ᐊᒻᓕᖅ, ᖅᑕᒍᔾ
ᐄᐊᑉ, ᑯᓗ ᐱᑦᓱᐃᖅ ᐊᒻᓗ ᓂᑎᓂᔾ ᓂᐅᑎᖅ, ᑕᐱᒃᑯᐊᖑᓕᖅ ᓂᐱᖏᑦ
ᓂᐱᓕᒥ ᑐᖕᓴᖅᑲᔪᓪᖃᐃ. ᐊᑕᖕᒧᔾ ᐱᒃᑕᐅᓂᒐᖅᓯᔾ, ᐅᑯᐊ ᒪᑉᐱᒑᑦ
ᑎᑎᕋᖕᓇᖅᑯᐅᓂᕐᖅᑦᒃᖅ ᐃᓃᓇᔨᓄᑦ ᐃᑯᔨᖅᑕᐅᓚᐅᖅᓃᒐᒪᐊᓕ. (I will forever be grateful to the Elders the interpreter Andrew Dialla and I interviewed in Pangnirtung, most of whom have since passed away. Thank you to Inuusiq Nashalik, Jamesie Mike, Pauloosie Veevee, Rosee Veevee, Evie Anilniliak, Daisy Dialla, Sowdloo Nakashuk, Peteroosie Karpik, and Elisapee Ishulutaq. I wish I could have met Etooangat Aksayuk, Pauloosie Angmarlik, Qatsu Evic, Koodloo Pitseolak, and Nowyook Nekootemoosee, whose voices I heard on tape. These pages could not have been written without all of you.)

I remain deeply indebted to the interpreter Andrew Dialla and his family. Andrew provided much guidance, teaching, help, and kindness beyond interpreting interviews and translating the above section. And while it takes uncommon skill to interpret two languages as different as English and Inuktitut, Andrew manages to translate both words and cultures with empathy and grace.

This book is grounded in a large number of written sources as well, and for these I recognize the staff of the Parks Canada Library; University of Calgary Library; Rutgers University Library; the New Bedford Whaling Museum; the G. W. Blunt Library at the Mystic Seaport Museum; the New London Historical Society; the Smithsonian Institution Archives; the Library of Congress

Manuscript Division; Library and Archives Canada; the Canadian Museum of History; the Providence Public Library; the Indian and Colonial Research Center in Old Mystic; the Houghton Library at Harvard University; the Rauner Special Collections at Dartmouth College; the National Museum of American History in Washington, DC; the Explorers' Club in New York City; the United States National Archives in College Park, Maryland, Washington, DC, and Waltham, Massachusetts; the Prince of Wales Heritage Centre in Yellowknife; the Danish National Archives in Copenhagen; the University of Alberta Archives; the General Synod Archives of the Anglican Church of Canada; the British Museum; and the Angmarlik Visitor Centre in Pangnirtung, Nunavut. Thank you to all the staff who helped me locate what I needed, and especially to those who brought me what I didn't know I needed. An extra thanks to Clare Flemming for the champagne, and to Robin Weber for knowing everyone in Yellowknife.

Financial support for research, networking, and writing was provided by the American Council of Learned Societies, the Mellon Foundation, the Enk Endowment Fund, the Network in Canadian History and Environment, the Social Sciences and Humanities Research Council of Canada, Rutgers University, and the National Science Foundation (grant no. 0715050). I am grateful to NSF program officer Anna Kerttula for her extraordinary thoughtfulness toward young scholars. Any opinions expressed in this book are my own and do not necessarily reflect the views of the funders or of my employer Parks Canada. An earlier version of chapter 1 was published in the conference proceedings *North by Degree: New Perspectives on Arctic Exploration*, edited by Susan Kaplan and Robert Peck.

I was lucky enough to visit many of the sites that I study. Everywhere I went, strangers, friends, and relatives welcomed me and showed me what they loved about their homes. They smiled at me, gave me chocolate and country food, took me to see places they loved, and let me stay with them. I would especially like to thank Oleepika Nashalik, Ooleepeeka Arnaqaq, Eeta Angnakak, Alukie Metuq and the late Noah Metuq, Hannah and Tina and Julia Tautuajuk, Shirra and Scott Baston, Carol and Charles McCollough, Lilian Oben, Patricia Pires, Maryse Saraux, and Pat Routledge and Will Scharbach. Also, thank you to everyone who generously gave their time and expertise to the University of Manitoba Pangnirtung Summer School for so many years.

There are many other people who have assisted in some way with the writing of this book over the past ten years. I could never list everyone, but thanks to all of you who helped me think and laugh, including Carmen and Ed Gitre, Tim Alves and Lesley Doig, Tina Adcock, Stacey Fritz, Kelly Enright, Leigh-Anne Francis, Justin and Stephanie Hart, Kenn Harper, Phil Goldring, Lyle

Dick, Michèle Therrien, Dawn Ruskai, Jan Lewis, Danielle McGuire and Adam Rosh, Louisa Rice and Eric Schanen, Rebecca Scales, Margaret Sumner, Amy Tims, Iben and Kristina Trino-Molenkamp, the SFU crew, and many others. For the best personal as well as professional role models anyone could ask for, thank you to Susan Schrepfer, Jackson and Karen Lears, Paul Clemens, Ann Fabian, Chris Trott, Paige Raibmon, Tina Loo, Charlene Porsild, Meg Stanley, David Neufeld, Frieda Klippenstein, and Joel Morassutti. I couldn't imagine better colleagues and partners than the ones I have at Parks Canada. It is an honor to work alongside so many knowledgeable and passionate people on projects that involve understanding place and history together. And although I disagree with the explorer Charles Francis Hall on many things, I can only concur with his 1863 statement, in a letter to Sarah Budington, that he "would rather make a dozen voyages to the regions of ice & snow than prepare one book for publication." I am deeply grateful to Tim Mennel, Rachel Kelly, Susan Olin, Yvonne Zipter, Jan Williams, and the staff at the University of Chicago Press for making this process as friendly and manageable as possible. I also sincerely thank everyone who advised and commented on portions of the manuscript, as well as the anonymous readers for their thoughtful and helpful critiques.

I could never have traveled so much without my own family to come back to. I learned some of the most important lessons about home from my grandfather Herb Routledge. My parents, Louise and Rick Routledge, and my sister Laura's family will always be a home to me. They fed me, reminded me of what was important, and made and kept me who I am. I also feel blessed to have the support of my American family—Ken, Diane, Cindy, and Yurii. And most of all, I thank Brian and Mira. I am grateful that we are able to make our home together in beautiful Treaty 7 territory. Brian, thank you for moving here with me, for always wanting to be outside, and for being the calm to my storm. Mira, my sweetheart, my little one, I love you all the time.

PROLOGUE

On the Ice

In October 1872, nineteen members of an American polar expedition became stranded on an ice floe. Their ship *Polaris* was squeezed or "nipped" in grinding ice off the northwest coast of Greenland. Expedition members clambered down onto the pack ice beside the leaking vessel, desperately throwing dogs, food, and supplies overboard. Beneath their feet the ice shifted and shattered. The *Polaris* broke away and was sucked into a night that was "fearfully dark" with intense blowing snow and sleet. The stranded crew used two small boats to gather themselves together. They were marooned on an ice island, on one of the broken pieces of frozen sea that scrape and groan their way southward every year.

The group was picked up six and a half months later by a sealing ship off the coast of Labrador. They had traveled fifteen hundred miles, drifting the entire way on pieces of ice. Their original floe had melted and broken apart, and they were using their one remaining boat to shuttle between ever-smaller floes. During high seas they were soaked by each large wave. Their dogs had perished. The crew was hungry, and some were feuding with each other. Yet all nineteen people had survived. In the words of the American survivor George Tyson, they had passed "through fearful perils and suffering, to safety, home, and friends."

There were two Inuit families among the expedition members on the floe; they were called *Esquimaux** in the reports of the day. They had been hired

---

* Today the term Inuit is sometimes used, especially in Canada, to refer collectively to Indigenous peoples once known to outsiders as Eskimos. Inuit is usually translated as "the people," containing an implication of "real people," although writer and translator Rachel Attituq Qitsualik argues that a more precise translation is "The Living Ones Who Are Here" (see Qitsualik, onto the *Polaris* expedition at its outset, and ended up largely responsible for the ice floe party's survival. The Inuit parents had five children between them; the youngest was just two months old when the drift began. The Inuit adults had been careful to take an adequate supply of guns and ammunition from the ship. They hunted, shared cooking and heating techniques, mended clothing, and supervised the building of ice shelters for the entire party.[1]

The Inuit fathers on the *Polaris* expedition had almost certainly been stranded on the ice while hunting before. Stories of ice floe drifts run deep into Inuit culture.[2] In 2008, I asked the Inuit Elder Inuusiq Nashalik if he had ever been stuck on an ice floe. He smiled and replied, "Many times . . . many times I've drifted out into the open ocean." Inuusiq explained that to return safely, he followed the advice of his father, Attagoyuk:

> My father always said, "If the piece of ice you're on breaks off, don't even bother looking at where you came from." . . . What my father told me to do was to take my dogteam and go to the other side, go to the other end of the floe. That is why you see me here today, because I always went to the other side of the floe, where that floe [would eventually touch] land or landfast ice. . . . My father used to always say, you have to be very patient. People have been lost because they had such a strong urge to get to safe ice.[3]

Inuusiq said that when cracks appeared in the ice next to the floe, that indicated the floe was pressed up against something solid. That was the time to dash to land.

Inuusiq was an expert on many things, including being interviewed by researchers. He would have understood that most outsiders find ice floe stories extraordinary. But the way he related them seemed quite matter-of-fact to me. Thanks to his father's advice and his own lifetime of experience, he had been able to deal with being stuck on the ice. He was patient but ready, observing the land and sea. He knew how to recognize "safe ice." Even the most skilled Inuit can die on the sea ice, but they are more often able to use their knowledge, preparedness, and ingenuity to avert disaster. Inuusiq's ice floe stories were part of being a hunter in his homeland.[4]

The *Polaris* ice floe drift is an Inuit story too, as there were Inuit present

---

"Is it 'Eskimo' or 'Inuit'?"). The singular form is "Inuk" and the dual (two) form "Inuuk." Inuit ancestral homelands are mostly above the treeline in eastern Siberia, Alaska, Canada, and Greenland. Regionally, Inuit or Eskimos self-identify in their own dialect or language. Today's Inuit Circumpolar Council represents Iñupiat in northern Alaska, Inuvialuit and Inuit in Canada's Arctic, Kalaallit in Greenland, and Yup'ik in Alaska and Chukotka (Russia). See the ICC's charter at http://www.inuitcircumpolar.com/icc-charter.html. Most of the indigenous people in this book came from Baffin Island and called themselves Inuit, so that is how I refer to them.

PROLOGUE

FIGURE P.1. Inuusiq, his wife Rhoda, and daughter Hannah ca. 1930s. NWT Archives/Archibald Fleming fonds/N-1979-050: 0731.

who later retold it. But it is not in the same category. Inuusiq never drifted fifteen hundred miles, for six months, with eighteen other mouths to feed. That experience was part of a late nineteenth-century culture of polar exploration, in which large numbers of outsiders tried to get as far north as possible. In short, both Inuusiq's experiences and the *Polaris* drift are narratives about humans on ice floes. But one is about the Arctic as home, while the other is usually retold as the opposite. Inuusiq's type of story is relatively unknown outside of Inuit communities, while the drift of the *Polaris* crew remains one of the best-known tales of polar exploration.[5]

It is also important to consider what happened after the *Polaris* crew was rescued. The first Inuit family chose to go home to Greenland. The father stated that he did not wish to settle in the United States. The second family moved to a New England whaling port where they had previously lived.

Within four years the young mother and daughter were buried next to each other in a Connecticut cemetery. They were neither the first nor the last Inuit to perish in a foreign country. In 1860, a man Americans called Kudlago had become deathly sick while visiting the United States. He was put on a whaling ship that spring, but died on the journey. In a reminder that the Arctic was his home, and that he longed to reach it, his last reported words were: "Do you see ice? Do you see ice?"[6]

This book is about ideas of home: how Inuit and Americans often experienced each other's countries as dangerous and inhospitable, how they tried to feel at home in unfamiliar places, and why this continues to matter. The Arctic has long been categorized by outsiders as a difficult place. In 1861, the future *Polaris* sailing master Sidney Budington summed up a common view when he reportedly asked Inuit on board his ship, "[Why] do you have such a cold, bleak, barren, monotonous, rocky, icy, stormy, freezing country here, unfit for a white man or any one else to live in?"[7] Adjectives like these settled over the Arctic. They shaped how future travelers, industries, and governments interacted with diverse northern landscapes and their inhabitants. They made it easier for Arctic locations to be valued mainly for their saleable resources, and for Arctic peoples to be seen as disadvantaged, inferior, and willing to relocate. In other words, they helped to justify colonialism. Budington's words contrast sharply with one way that Inuit in that same region today describe their land—as *nunattiavak,* meaning "good land" or "beautiful land." The translator Andrew Dialla explained that *nunattiavak* is said of a place that has access to animals, a level place for tents or homes, and usually a good harbor.[8] While Budington described everything around him as "monotonous," Inuit see specific, familiar, life-sustaining places linked together by travel routes. Today the phrases "Arctic homeland" and "Inuit homeland" have become commonplace in popular and scholarly literature, and Inuit have long been advocating for their own interests in their homelands and beyond. Yet enduring assumptions about the Arctic as a grim and inferior region still hold too much power.[9]

In this book, I return to the time when stereotypes like the ones that Budington itemized were taking hold. Arctic exploration narratives, including the *Polaris* crew's drift on the ice floe, became extraordinarily popular in nineteenth-century North American and European cities. Many earlier exploration stories and sketches featured Inuit at home and contained other domestic or picturesque scenes—for example, sailors playing cricket on the sea ice. But by 1860, with the disappearance of British explorer Sir John Franklin and his entire crew, the Arctic was increasingly written, painted, and imagined as a bleak, harsh, monolithic expanse covered in snow.[10] Some exploration accounts were nuanced and positive, but the popular press preferred more sensational stories of heroic men crossing icy wastelands, their lonely ships crushed in the ice. These stories peppered books and newspapers, paintings and lantern slides, lecture halls and music halls. They came to represent "the Arctic" for many people who never traveled there. But what lies beneath their surface is much larger and deeper. Most people who lived in or visited the North American Arctic in the nineteenth century did not call themselves explorers.[11]

This book considers stories of Inuit and Americans at home and away from the 1850s through the early 1920s, to examine how American Arctic stereotypes were constructed as well as what they missed. These stories were originally told in two main languages: English and the Inuit language of Inuktitut.[12] I use the Inuktitut term *Qallunaat* to refer to non-Indigenous outsiders—mostly Americans, Canadians, and Europeans who would today be considered "white."[13] Inuit and Qallunaat had—and many still have—very different ways of seeing and imagining the world. To give just one example, Inuusiq Nashalik also spoke to me about the importance of some stone markers, or *inuksuit,* that his father had shown him. Andrew Dialla, a highly skilled interpreter, struggled to express Inuusiq's Inuktitut statements in English. First, Dialla translated that these *inuksuit* "point all the way out into the land, way out into the middle of nowhere." Then he paused and reconsidered: ". . . but into *somewhere*, because they always pointed to a place where there's usually caribou."[14] In English, the Arctic is often the "middle of nowhere." In Inuktitut, relationships to animals and the land define specific places as "somewhere." The tension between the "middle of nowhere" and the "middle of somewhere"—between Qallunaat and Inuit visions of the Arctic and how to live there—is at the center of this book.

Underlying all of these stories is a common thread of colonialism. The terms "home" and "homeland" are political. Americans built their own homeland through the dispossession of Indigenous peoples. This was a violent backdrop—and in some cases foreground—to the lives of Americans in this book, as the United States pushed westward, fought wars, and forced Indigenous peoples onto reservations.[15] Americans never seriously tried to annex the eastern Arctic—Britain and later Canada would do that. But American incursions there were part of a quest for power and resources. Shiploads of Americans arrived in Inuit homelands, where they often behaved as though the land and sea belonged to them. The stories in this book are about the unmaking and remaking of homes that is the beating heart of colonialism. I see colonialism here as warp thread in a tapestry, what Audra Simpson describes as an "ongoing stress and structure" that Indigenous people endure and push against. Colonialism was, and remains, incomplete and resisted.[16] It is vital to understand what it does and continues to do, but at the same time, to see how Inuit and Qallunaat interwove their lives—and kept them apart—in ways that can defy prediction.

*

"The Arctic" is a vast region, often a mythical, undifferentiated mass of snow and ice to those who live outside of it.[17] To avoid generalizing about northern landscapes, I have anchored this book around a single place: Cumberland Sound. Cumberland Sound is a large bay on Baffin Island that faces out toward the Davis Strait and Greenland. It is just south of the Arctic Circle, the latitude above which the sun never rises or sets at least one day per year. Today Cumberland Sound is in Nunavut, a Canadian territory established in 1999 under a treaty between Inuit in this region and the Canadian state.[18] Dotted with islands and deep fjords, Cumberland Sound is bounded by granite outcroppings and rolling hills. A wall of jagged glaciated peaks is visible to the northeast on clear days. Cumberland Sound has a variety of Inuit names, including Tinugivik and Ikaraq. In the nineteenth century it was also known in English as Cumberland Straits, Cumberland Inlet, Cumberland Gulf, Northumberland Inlet, and Hogarth's Sound.[19] Many of the people in this book lived or worked here. The sailing master and assistant navigator of the *Polaris* expedition both worked on whaling ships in Cumberland Sound. One of the Inuit couples from the ice floe drift was born in the region. So was Kudlago, the Inuk who died on his way home, asking about ice.

Cumberland Sound is Inuit territory. Approximately eight hundred years ago, the ancestors of today's Inuit, known to archaeologists as the Thule culture, left what is now Alaska. They migrated—likely within the timespan of a single generation—across the Arctic to northwest Greenland.[20] Some of their descendants settled in Cumberland Sound. Inuit relied on the products of whales, seals, caribou, walrus, and other animals for their food, clothing, fuel, tools, and shelter. In Cumberland Sound, marine mammals and especially ringed seals were the dietary staple, and Inuit burned blubber in lamps for heating and cooking. Families traveled long distances by dog team, on foot, and in boats. This mobility does not imply a lack of permanence, but rather a permanence of shifting patterns, of animals and names and communities and seasons. In many locations around Cumberland Sound today, remains of old dwellings, or *qammat*, are easily seen. These attest to the history of people returning over and over again to the same sites.

The English navigator John Davis visited Cumberland Sound briefly in 1585 and 1587, but the area remained relatively unknown to outsiders until 1840. That year, a young Inuk named Inuluapik led the Scottish whaling captain Wil-

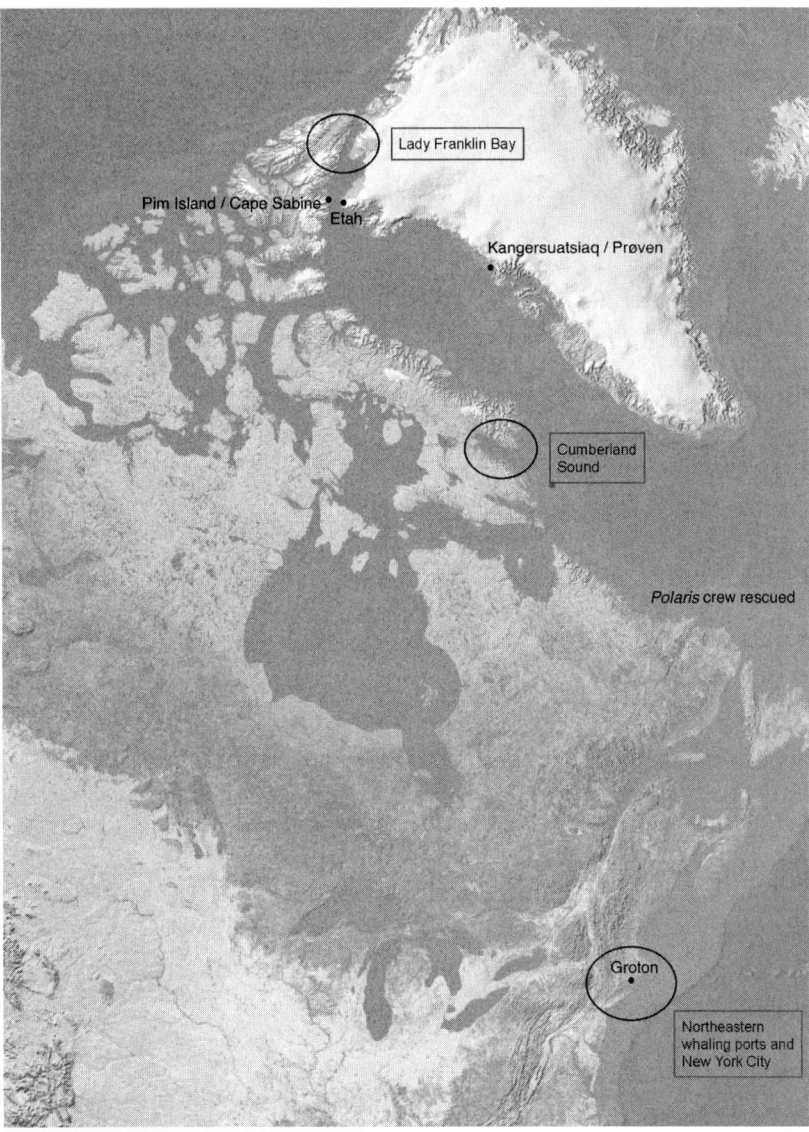

FIGURE P.2. Overview of Cumberland Sound and other key regions in the book. The *Polaris* crew's drift on the ice floe began near Etah, and the Qillarsuaq voyage ended there. Background map courtesy of the United States National Park Service.

liam Penny into his home territory and sparked a Qallunaat rush on bowhead whales.[21] Cumberland Sound became a last gasp of the staggeringly profitable Davis Strait whaling industry, in which, at best estimate, over 28,000 bowhead whales were shipped out between 1719 and 1911. Several thousand of these whales probably came from Cumberland Sound.[22] Qallunaat whaling crews,

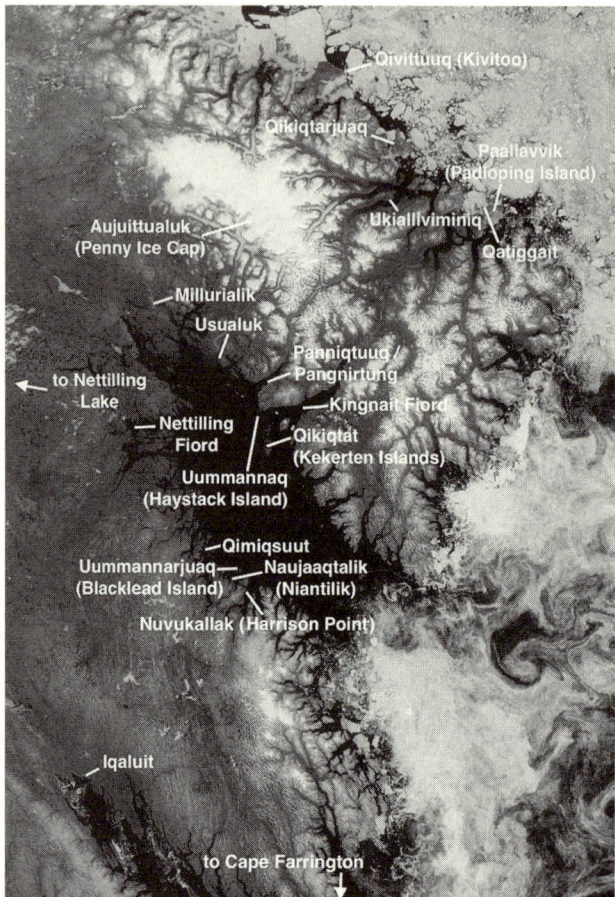

FIGURE P.3. Sites in the Cumberland Sound region mentioned in the text. Inuktitut place names are from Inuit Heritage Trust, which has documented hundreds of additional place names in this region; see ihti.ca. Qallunaat whaling names are from Goldring, "Whaling-era toponymy." Aerial photo credit: NASA's Earth Observatory, July 11, 2009.

mostly working for Scottish and American companies, began to overwinter in Cumberland Sound in the 1850s. They froze their ships into the ice and built stations to house employees and process blubber into oil. Cumberland Sound quickly transformed into one of the intensive resource-extraction job sites that continue to be a part of the Arctic world, with most of the profits flowing south.[23] In addition to the Inuit who already lived there, others migrated in from hundreds of miles away to work and trade. Inuit in this region were experienced subsistence whalers. Some of their ancestors had likely traded baleen in Hudson Strait in the 1700s. By the 1870s, Inuit made up most of the whaling workforce in Cumberland Sound.[24]

PROLOGUE   xix

The whaling grounds began to attract scientists, missionaries, and researchers. A group of German scientists set up a research station during the First International Polar Year in 1882–83, and the anthropologist Franz Boas did his first fieldwork in Cumberland Sound in 1883–84.²⁵ By this time, the bowhead whale population had severely declined, and Inuit had diversified into trading furs and other products with visiting ships. The bowhead whaling industry ended in the early twentieth century. In 1921 the Hudson's Bay Company moved in and established a trading monopoly. They encouraged fox trapping for the fur trade. In the 1960s, the Canadian government put increasing pressure on Inuit across northern Canada to resettle into centralized permanent townsites. Today most local Inuit live in the hamlet of Pangnirtung. This relocation is discussed in more detail later, but it dramatically altered Inuit life and their relationships with Qallunaat.²⁶ Due to the cost of gas and snowmobiles, low sealskin prices, and work or school schedules, getting out on the land is no longer simple for many people. Nevertheless, many Inuit are still subsistence whalers; and they continue to travel, hunt, and camp out on Cumberland Sound.

During the commercial whaling period, people from many nations passed into and through Cumberland Sound. They mined a wealth of commodities and ideas from the margins of their known world. Cumberland Sound whale oil lit cities and greased trains; baleen corsets shaped fashion and femininity. Seal and fox and polar bear furs, ivory and antler and soapstone carvings— all made their way from Cumberland Sound onto Victorian Qallunaat bodies and into the parlors of far-flung homes. Many Qallunaat defined their ideas of home against what they imagined the Arctic to be. Through all of this, Cumberland Sound remained and remains an Inuit homeland that, for centuries, has existed at the center of itself.

Cumberland Sound is at the center of this book, but I follow people into and out of it. Chapter 1 is about Americans in an Inuit homeland, working in the booming commercial whaling industry in Cumberland Sound in the 1850s and 1860s. Chapter 2 deals with Inuit in an American homeland, focusing on the couple from Cumberland Sound who ended up on the ice floe. They are renowned for their six-month drift and for their expeditions with explorer Charles Francis Hall, but I discuss their lesser-known years in the United States. Chapter 3 features Inuit and Americans traveling together to a remote polar site where they encountered no other people. I focus on the experiences of an American enlisted man and a Kalaaleq (West Greenlandic Inuk) employee on the disastrous US government expedition to Lady Franklin Bay, on which nineteen men perished. Chapter 4 discusses Inuit who remained in their homeland as it changed around them. It returns to Cumberland Sound and asks how people there remembered the early twentieth century when commercial whaling was winding down, and the whales were all but gone.

I look at Inuit and Americans at home and away because human knowledge of place is built not just on local experiences and stories, but also through journeys and in relation to other regions. As geographer Yi-Fu Tuan writes about home, "Implied is something that is not home and that lies outside of it."[27] Focusing on Cumberland Sound while comparing it to other sites allows for a fuller understanding of the limited nature of Qallunaat stereotypes about the Arctic.

*

What does it mean to be at home? Home for humans is much more than just shelter. It is the shifting and emotional connections among the people, places, and ideas most central to us. Our sense of home is shaped by relations within our own communities, but also with outsiders. Some people have strong attachments to several home-places simultaneously. Feelings about home are often positive, but they can also be negative, ambiguous, or troubled.[28] In the words of writer Verlyn Klinkenborg, home is "a place we can never see with a stranger's eyes for more than a moment."[29]

The ways in which Americans and Inuit thought about home were diverse and complicated, and I could never claim to know them all. The English word "home" merits an *Oxford English Dictionary* definition that is one-third the length of this book. In Inuktitut there are many words related to home or feelings of home, but the most common translation for home is *angirraq*. Today Inuit use the term *angirraq* in an everyday sense of going home or being at home. It is also the root of an expression for homesickness, *angirraqsiqtuq*, and it is used in translations of imported concepts like homelessness, *angirraqanngitut*. Like the English word "home," *angirraq* also has abstract and emotional connotations. In the Igloolik area in the twentieth century, the late Elder Michel Kupaaq and others used the term *angirrattinni*, "in their own homes," to refer to places out on the land where Inuit used to live. The same hunters described returning to the hamlet where they were resettled by the Canadian government as *qallunaajjariaq*, "going to the Qallunaat." They continued to do this for at least fifteen years after moving into town in the 1960s.[30]

At least one important use of the root *angirraq* has no correlation in English: it relates to people's names. An Inuit name, or *atiq*, is not just a name in the English sense. It also contains social relationships and personality attributes. Generations of human bodies pass through Inuit names, and in most Inuit homelands all names can be held by both men and women. After the name of a deceased Elder enters into a baby or a fetus, people address the child with kinship terms they used for the deceased person—so a girl might be called "grandfather" by her biological mother. Anthropologist Christopher Trott pictures Inuit names as a permanent grid over a homeland, which people pass through in their lifetimes. The association between names, land, and home is strong. At least in some dialects of Inuktitut, bringing the name back is expressed as the name returning home. According to tradition in Cumberland Sound, a person should be dead for a year before the name returns, although this is not always practiced today. Names were usually bestowed after people began dreaming of the namesake; this was an indication that the deceased wanted the baby named after him or her. In the words of Kivalliq Elder Salome Qalasiq: "When my daughters and I started dreaming about my deceased son, we gave his name to a child because we kept on dreaming about him coming home."[31] Passing on names not only ensures the name will continue to live on the land; it also builds networks of home and belonging that extend far beyond the nuclear family or biological relatives.

I do not fully understand how nineteenth-century Inuit families in Cumberland Sound thought about home. But a sense of a communal past continuing into the present and on into the future seems to have been crucial, with naming being one aspect of this. Home was where your people had lived before you were born, and where they would continue to live. It was where those who shared your name had traveled the same routes. Some Inuit moved away and formed strong attachments to new places. But as the anthropologist Hugh Brody has noted, hunter-gatherer societies are generally less nomadic than agrarian ones. Hunter-gatherers have often lived in the same territory since time immemorial, whereas some farming peoples are always pushing out, looking for new land and resources.[32]

The contemporary Inuk writer Rachel Qitsualik states that, "Just as one knows one's neighborhood, or the town where one was raised, so Inuit regarded the entirety of the land as their home."[33] Similarly, geographer Béatrice Collignon suggests that Inuit she worked with in Ulukhaktok in the late twentieth century had a sense of place that was less about having a fixed "place of their own," like the homestead or private house that was a marker of success and class in American society. Instead, they focused on continuing to access a variety of places with diverse resources and different animal populations. These locations were familiar, connected by trails and travel routes, and held together by place names and the history of the community. This is a sense of home that extends over a very wide region.[34]

The commercial whaling period was likely experienced as a time of upheaval. Some Inuit families moved across hundreds of miles to contact foreign whalers.[35] Cumberland Sound was within their larger homeland, and probably within their network of names and relationships. But they would have left their birthplaces and core hunting areas to live by whaling ships and stations, alongside Qallunaat and Inuit with whom they had at first only distant connections. Many suffered and died from new diseases. For the Inuit, bowhead whales were a source of food; their oil was used for heating; their ribs framed homes; their jawbones were used for sled runners. And Qallunaat companies shipped nearly all the whales out. People missed the whales and missed the connection they had to them. Their loss affected where Inuit lived, how they traveled, and the tastes and smells inside their dwellings. Still, the land remained home.

Most Americans in this book had less steadfast and ancestral attachments to a single region than did Inuit. They were new immigrants, or descendants of settlers and slaves. This does not mean they had no sense of home in the United States. By this time, some settler families had occupied the same land for centuries. One woman, Lucy Larcom, wrote of her hometown in Massachusetts in 1889: "There is something in the place where we were born that holds us always by the heartstrings.... [A town] rich in beautiful scenery and ancestral associations, is almost like a living being, with a body and a soul. We speak of such a town, if our birthplace, as of a mother ... its miles of sea-border, almost every sunny cove and rocky headline of which was part of some near relative's homestead."

Lucy Larcom left her hometown as a child after her father died. Her older brother went to sea, and her widowed mother moved inland to open a boarding-house for girls and young women who worked at the mills in Lowell. By the age of eleven Lucy was working at a mill herself.[36] In the nineteenth century, millions of Americans like the Larcoms uprooted themselves to seek better economic opportunities for themselves and their loved ones. Young people set off for mills and factories, farmers moved into cities, settlers and miners fanned out across the West, and whalers and merchants sailed off on ever-longer voyages to exploit resources around the world. They disrupted other people's homes, often violently. A few Americans turned vast profits, as evidenced in the whaling town of New Bedford, Massachusetts, where ships arrived heavily laden with oil and baleen from distant whaling grounds. Novelist Herman Melville described the rows of lavish mansions as being "harpooned and dragged up hither from the bottom of the sea."[37]

American ideas about home were changing in this period, first among the middle-class and later more broadly. People began placing more importance on their houses and immediate family over the larger land and community. This fit well with a society on the move, since a house could often be refilled with the same furniture and relatives in a new place. The ideal house became a private, isolated sanctuary for a husband, wife, and their children. Work, education, and care for those outside the family circle were now supposed to take place elsewhere. This ideal was never reality. If working-class families owned a house at all, they likely took in paying lodgers and other people's laundry, housed newly arrived relatives, converted kitchens into bedrooms at night. Husbands went to sea in search of work. Wives and children labored in factories; and of course, domestic labor took place inside all homes, even if it was not considered "real work." But still, middle-class homes became more isolated from the community. And this ideal of home was powerful enough to put young men on whaling ships, in the hopes of making enough to marry and buy property. Most never became wealthy, but the quest to drag mansions "up hither from the bottom of the sea" reflects a tension in how Qallunaat have thought about home since industrialization: as people come to expect and desire a higher standard of living, leaving home is often necessary to make a home.[38]

By the mid-nineteenth century, white middle-class American men were encouraged to view attachment to home and place as dispensable, and to move in the search for progress and wealth. Some thrived. But others—including many in this book—missed home terribly and longed to return.[39] Overall, mass mobility in the nineteenth century had two opposing results for how Qallunaat thought about home: it served to decouple place from home, and it produced a wave of homesick Americans who spent a great deal of time thinking about home and place.

The Inuit and Americans in this book had strikingly different worldviews as well as home landscapes. Animals, family, gender, children, individuals, the afterlife, the human place in nature—many of these concepts could barely be translated between English and Inuktitut. But some basic concepts of home seem to have crossed cultures. A good home was a place, or a network of places, where people felt secure; where their ways of thinking, being, and living made sense; and where they believed they and their loved ones had a future. A sense of belonging was important too, as well as responsibilities and strong relationships, and feeling a part of something larger. Not everyone had an equal opportunity to obtain this ideal of home, and of course the specifics varied widely: who constituted family, or the sounds and smells that should fill a good house, or the values and skills most important to pass onto children. But when the people in this book talk about "home," they often mention family and friends and children in the same breath. They also speak of familiar weather, landscapes, food, clothing, shelter, heating, and modes of travel. Seasons, holidays, and other ways of marking time are similarly important.

✳

I want to briefly address how I see this book fitting into the academic literature on Arctic history, and where I feel it can suggest new directions, especially for environmental history and histories of home. First, I draw inspiration from the work of scholars like Emilie Cameron, Julie Cruikshank, Andrew Stuhl, and Lyle Dick, and the histories produced by the Qikiqtani Truth Commission. These works have all considered how Indigenous and Qallunaat knowledges and histories intersect in what is now northern Canada and Alaska.[40] Far more writing focuses on either Qallunaat or Inuit lives, and while this is appropriate for some topics, there is much to be gained by looking at these histories together. Inuit have been living with Qallunaat for over 150 years, and have plenty of insights into Qallunaat culture and ideas about home. Inuit today, in Canada or the United States or Russia or Greenland, are forced to think about their history in relation to outsiders all the time, in a way that most Qallunaat are not. Qallunaat-Inuit exchanges have been marked by enduring inequities, exploitation, mistakes, and misunderstandings; but they were and are not monolithic or unchanging. Paradoxically, by bringing Inuit and Qallunaat histories together, I also show how they are not fully connected. Even when Inuit and Qallunaat worked side-by-side, their lives had meanings and joys and crises unknown to others.[41] Inuit of course have far more to say about this and other topics than I know or can convey in this book. I am trying to show what I have learned so far, with the recognition that it is incomplete, not definitive, and open to correction and reinterpretation. I encourage non-Inuit readers to seek out Inuit perspectives and learn to listen to them, as I will continue to do.[42]

This book is fundamentally about Inuit and Qallunaat trying to feel at home in changed or unfamiliar places. I first came to think about the importance of place in history through the work of environmental historians, who study the ways humans have related to nature, been a part of it, changed it and been changed by it.[43] Some environmental historians have considered how people have acclimated to new surroundings when they moved or others transformed their homelands.[44] These studies are useful here, but environmental historians have written surprisingly little about ideas of home. Perhaps this is because academic histories of home have traditionally focused on the household: family relationships, women's work, racialized gender roles, child-rearing, and domestic life.[45] In much academic writing, settlers have "homes," but Indigenous people have "homelands." There is some truth to this dichotomy, but it reflects Qallunaat ideas too uncritically. As we have seen, urban Qallunaat Americans did come to place a high value on their houses and the nuclear family, while Inuit kept the land and relationships at the center. But homesick Qallunaat missed more than just their house and immediate family, and Inuit who were relocated into wooden houses reacted to the change of architecture as well as location. Today two pressing issues in Nunavut are the lack of adequate housing and continued access to traditional lands. If we understood everyone's home as including shelter but extending far beyond it, what more could environmental historians say about home? About how Qallunaat ideas of home have impacted treatment of other peoples and their land? Or about the relationship between home and environment, between home and environmental change? Home is a key point where nature and culture meet; it is a topic to which environmental historians could bring new insights.

Similarly, much attention is given to space and place in studies of home, especially by geographers, but far less to time.[46] Yet our homes are where we most clearly mark the regular passage of time: waking, sleeping, eating, dreaming about the past and future, watching children grow up and loved ones grow old. When people leave their home for somewhere very different, they are often forced to question their fundamental ways of seeing and ordering the world. Their previous understandings of time can fail them. Even ecological markers like seasons, the sun, the stars, the moon, and the tides are not the same in new places. Each chapter of this book traces a way of marking time that was central to that particular environment: six Inuit seasons in Cumberland Sound, a series of Sundays in the northeastern United States, and the path of the sun in the High Arctic, with its annual appearances and disappearances. Time was not the only factor impacting people's sense of home, but this chapter structure highlights time's role.

Thinking seriously about time adds another dimension to our understanding of Inuit and Americans at home and away. Americans and Inuit imagined and understood time in a myriad of abstract ways in the nineteenth century. Inuit had a sense of mythic time that linked people to the very beginnings of human life, as well as *longue durée* ecological factors like shifting animal migration routes.[47] Most Americans situated themselves in various permutations of sacred time. The idea of heaven as the ultimate eternal "home" was popular among American Christians, free and enslaved, whose families lived apart on earth but looked forward to being together again after death.[48] Sacred time was also marked in this world with holy feasts, Sabbath rituals, and seasonal holidays—events that Americans frequently celebrated and wrote about when they were away.

For both Inuit and Americans, concepts of time shifted in the nineteenth century as their homes and surroundings changed. Many Americans adopted the industrial work week and acquired personal clocks. The Second Great Awakening and later revivals transformed the way countless people thought about religion. Inuit adapted their origin myths and rituals to deal with the changes brought by commercial whaling, and they learned the concept of weeks from whalers and missionaries. As Inuit in Cumberland Sound converted to Christianity in the early twentieth century, they entered more deeply into Christian conceptions of time, including an eschatological sense of movement toward the end of days.[49]

Colonial encounters also shaped how people thought about their place in history. Victorians frequently expressed a sense of "stepping back in time" in the Arctic. Some Qallunaat believed that "the Esquimaux are . . . the 'survivals' of the Cave Men of Europe," meaning that they were early humans who had been pushed further and further out by the state, until they were "pressed within the Arctic regions." By this logic, the Arctic was a place where people would never choose to live, where they became trapped or "pressed within" the past.[50] Of course this was not true; Inuit lived in the Arctic by choice, and they had been developing technologies and refining knowledge alongside all other humans. Presumably nineteenth-century Inuit did not perceive themselves as throwbacks in the Qallunaat narrative of human progress. But I do not know how exactly they situated themselves in history, and to what extent they experienced the arrival of the whalers and their diseases and trade goods as a break with the past, or as repeated rupture, part of a tradition of change.[51]

★

Before moving on, let's briefly return to the American expedition ship *Polaris*. When this vessel broke away in October 1872 and left nineteen people adrift on the ice, fourteen crewmembers were still on board the ship. They were the fortunate ones. The captain ran the leaking ship aground on Greenland, near the Inuit settlement of Etah, where the people were fascinated by the ship's cat, as they had reportedly never seen a cat before.[52] The following summer, the *Polaris* crew headed south in boats and were picked up by a whaling ship. During their winter in Greenland, they met an Inuk who spoke English well. Her name was Ivalu, and she had a tattooed face—common practice among some Inuit at the time, but not at Etah. She told the captain that she and several other families had arrived a few years earlier; some of them had come all the way from Cumberland Sound. Led by an influential shaman named Qillarsuaq, they had traveled over fifteen hundred miles on foot, in small boats, and by dogsled.[53]

Qillarsuaq, the shaman, was said to have murdered a man on a caribou hunting trip—unusual behavior for Inuit of his time, most of whom avoided overt displays of violence and anger. Qillarsuaq and his accomplice Uqi fled north to escape retribution. Several dozen family members and followers came with them. They traveled for hundreds of miles to the northern coast of

PROLOGUE   xxvii

Baffin Island, but there they made new enemies. After being ambushed and climbing atop an iceberg to pelt attackers with arrows, Qillarsuaq decided to flee again. The group waited that winter for the sea ice to "almost stop moving" and then crossed to Tallurutit (Devon Island) by dogsled.[54] Their sleds were unusually large; they were pulled by up to twenty dogs each and could carry kayaks. Recent whaling shipwrecks had likely offered up this uncommon abundance of wood.[55]

Although it was not unheard of for Inuit to visit this island, there were probably no other settlements there when Qillarsuaq arrived. The party spent several years at peace, separating into smaller camps to better harvest diffuse resources. In 1853 and 1858, Qillarsuaq and some of his followers met ships searching for the missing British explorer John Franklin. The second captain, Francis McClintock, recorded that the group of Inuit were "very fat and healthy." Still Qillarsuaq could not rest. He wanted to reach Inuit he believed were in northern Greenland. He made "spirit flights" through the air, assessing the route and watching the strangers. He once again convinced the full group of about fifty people to join him.

According to Merqusaq, a young man on the voyage who was Qillarsuaq's sister's son, the party found "plenty of animals" and stayed well fed through the journey to Greenland.[56] This time, however, the center could not hold. Qillarsuaq's old accomplice Uqi began to accuse him of lying about his spirit journeys, and he encouraged others to turn back. Merqusaq reportedly said that Uqi "had long been grave and without words, and then all of a sudden he began to talk about whale-beef. He was homesick for his own country, and he wanted to eat whale-beef again."[57] Bowhead whale was a common food in Cumberland Sound, and when the whale population crashed due to commercial whaling, many twentieth-century Inuit would long for it too. Finally Uqi turned around and began the long journey home, taking half the group with him. Qillarsuaq's group continued on a short way, stopping at Pim Island. There Qillarsuaq and his followers crossed the frozen strait to Greenland.

When they met local hunters there, the two groups excitedly exchanged news and struggled to understand each other's dialects and customs. The Baffin Island Inuit proved very helpful. Some years earlier, following a devastating epidemic, the Etah community had lost their knowledge of how to build kayaks. Without these boats, their homeland was a smaller place. They were landfast when the sea ice melted each year. They also lacked bows and arrows, fish hooks and spears. The newcomers reintroduced these technologies, making it far easier to procure a wider variety of game year-round.

After keeping himself in check for several years, Qillarsuaq again lost his temper and killed a man. From then on, he suffered stomach pains, which he and others believed were brought on by "the man he had killed [getting] inside him."⁵⁸ Qillarsuaq also longed for home. He "wished to see his own country again before he died."⁵⁹ For the last time, he convinced most of his family to follow him and return to Baffin Island.

Qillarsuaq died on the trip home. The party was still hundreds of miles from their destination. The game disappeared, and so did one of the group's mothers, Patdloq. She left to pick up a bowl she had left at the previous campsite, carrying a child on her back. When she did not return, people traced her footsteps, which became further and further apart. They had to search carefully for the next print, and finally all trace of her vanished. The others believed that she had risen into the air; some speculated that she might have left to avoid what was coming next. Members of the party soon descended into insanity, murder, and cannibalism. Five survivors turned back toward Greenland on the verge of starvation. One of them was a small girl, Ittussaarsuaq, who would later tell this story. She struggled behind the others in her father's heavy warm coat, often stumbling and falling behind. "Just like a dog that was starving! But she was starving too," recounted Inuuterssuaq Ulloriaq, who wrote down her tale.

Ittussaarsuaq's family waited for her to catch up. Dragging their possessions because they had eaten their dogs, they arrived back at Etah five years after they had left. They made it their home, and their descendants still live in the region today. But they did not forget the old country. Near the end of his life, the survivor Merqusaq reportedly said, "I traveled so far that I never made it home again. You should not do this, for it is not good to long for home when one gets old."⁶⁰ In the 1970s, one old man in Greenland asked before he died for his name to "be returned to its proper land." The name was brought home and given to a child in Ikpiarjuk or Arctic Bay.⁶¹

This story is an Inuit one, recorded in pieces by others but passed down and kept alive in Inuit homes. It has fascinated outsiders, arguably in part because of its commonalities with Qallunaat polar exploration stories. Geographically it is nearly the *Polaris* ice floe drift in reverse, and the Qillarsuaq migration is also shaped by food shortages and hardships and powerful—perhaps power-hungry—men on the move. Yet it is also a strikingly different tale, of intentional movement, unrolling over decades, and everywhere suffused with supernatural occurrences as common and accepted as bodies or breath. Qallunaat appear at the edges, but they are not the central characters making the decisions. The story deals with wide-ranging Inuit community and kinship, and the perils of human isolation. It reflects a central concern of this book: what can happen when people leave home, look for home, and try to return home?

1

Americans in Cumberland Sound

July 3, 1860, was a calm day in the Arctic waters of Cumberland Sound. The American whaling bark *Antelope* slid slowly along, cruising for bowhead whales. The ship cut through lifting fog and passed drifting icebergs that glowed blue and white against the dark water. Hills and rocky islands faded in and out of view. On deck, the ship's officers were attempting to deal with an insubordinate, low-ranking crewmember recorded only as Peter. Peter had been told to climb up into the riggings, probably for his regular watch, but he refused to go aloft. Instead he jumped out onto the ice beside the ship and shouted that he "would not come aboard again." The third mate eventually clambered off the ship and dragged Peter back on board, presumably with the entire crew watching. The *Antelope*'s logbook offers no clues to what Peter hoped to accomplish with this brief act of defiance.[1]

In 1860, bleak Arctic stereotypes were gaining widespread currency. Icebergs, polar bears, and Arctic sailors adorned buttons, dishes, playing cards, tea tins, and textiles. The previous year, the press had reported evidence that the British explorer Sir John Franklin, and likely his entire crew of 128 men, had perished while seeking a northern sea route to Asia that had become known as the Northwest Passage. The mystery of Franklin's disappearance had gripped British and North American publics for a decade. In the United States, the explorer Elisha Kent Kane had captivated a broad audience with his searches for the missing men. The press coverage of Franklin, at least in Britain, was at its most intense in 1859–60. Popular ideas of the Arctic as a harsh and desperate place were hardening.[2] Peter's decision to abandon the relative safety of an Arctic whaling ship—for a piece of ice, no less—would have struck many of his contemporaries as suicidal.

But what if Peter actually wanted to be left on the ice? The environment into which he jumped bore little resemblance to Arctic stereotypes. In early July, it was the Inuit season of *upingaaq*: the warmest, wettest, most verdant time of the year. In this season it was never dark, not even at midnight. Cumberland Sound could be drizzly, chilly, and damp in *upingaaq*, but it could also be warm and sunny. On the day Peter jumped ship, the *Antelope*'s logbook keeper described the weather as "pleasant."[3] In 1860, there was still pack ice moving out of the sound on its way south, but most of the area was easily navigable. Many of the rocky hillsides were carpeted in greenery and flowers. Waters and shorelines were home to whales, seals, caribou, Arctic char, clams, and a variety of migratory birds.

Even if Peter's captain had abandoned him, he could have turned to others for assistance. Hundreds of Inuit inhabited Cumberland Sound in the mid-nineteenth century.[4] Inuit families approached arriving whaling ships in boats, bringing furs and fish to trade. One captain recalled that on arrival, his "cabin was soon crowded [with Inuit visitors], and not only the cabin, but the cabin-steps, the companion-way, and the after-part of the deck."[5] There were also dozens of Qallunaat in the vicinity. Both American and Scottish whaling ships visited Cumberland Sound that spring, and several of them were in sight when Peter took to the ice. Even at the height of the search for Franklin, there were ten times as many commercial whaling ships as expedition vessels in Arctic waters.[6] These ships frequently exchanged equipment, provisions, letters, and employees with each other. That winter, along with eleven other Scottish and American vessels, the *Antelope*'s captain would anchor next to Inuit communities and deliberately freeze his ship into the ice.[7] In short, Cumberland Sound was not a deserted place.

But if Peter did want to abandon ship, he was in the minority. Most Qallunaat whalers struggled to see Cumberland Sound as anything approximating home. Many were miserable, fell ill, and perished there. I will consider why these whalers legitimately experienced Cumberland Sound as a hard place to live, how this related more to their background and circumstances than to an inherently harsh environment, and what this can tell us about how Qallunaat have related to the Arctic and to Inuit.

American whalers who kept logbooks and journals in Cumberland Sound often considered winter to be the only season worth noting. They contrasted Arctic weather with their own seasons, and the climate constantly reminded them they were not at home. Yet at least six seasons are self-evident to many Inuit, just as four seasons are perceived in most temperate regions and three in many tropical ones. Below I follow the overwintering whalers through six Cumberland Sound seasons: *aujaq* (summer), *ukiaksaaq* (early fall), *ukiaq* (fall), *ukiuq* (winter), *upingaksaaq* (early spring), and *upingaaq* (spring). These are not the only possible seasons. Seasonal distinctions are somewhat arbitrary in all societies, and some Inuit consider there to be more than six seasons or use more specific seasonal terms depending on the context of the conversation.[8] The timing of the seasons also varies from year to year, as they are tied to the weather and movements of animals. Even though American whalers did not use these Inuit seasonal terms, their records show distinct events occurring as each season turned.

Aujaq (Summer)

Peter was not the first whaler to abandon ship in Cumberland Sound. Nine years previously, on September 1, 1851, the American vessel *McLellan* sailed into the sound. It was then *aujaq,* or summer, the season when the waters are freest of ice. The air was getting drier and crisper. The sun dipped below the horizon, and in August it had become dark enough for stars. Berries were ripening; the leaves of some of the plants that clung to the shorelines had begun to change color. Cumberland Sound Inuit harvested the downy seedpods of the Arctic willow in this season; when mixed with dried moss, these provided the wicks for the stone oil lamps that were the source of light in their winter houses.[9] They also traveled inland to hunt caribou, primarily to get the skins necessary for winter clothing. During *aujaq,* the whales began to migrate out of the sound, but sightings kept whalers busy and hopeful throughout the season.

In 1851, the crew of the *McLellan* caught "a few whales," but not enough to fill their ship. A local man told the crew that whales were most numerous earlier in the year, when the entrance to the sound was still too choked with ice for foreign ships to enter it. The only way to hunt these whales would be to spend the winter in the sound, in order to be there when the whales returned. No Qallunaaq had ever before attempted to overwinter in Cumberland Sound, but Arctic explorers had done it elsewhere.[10] Captain William Quayle asked if twelve men—a third of the crew—would consider spending the winter, probably near the Inuit community of Qimiqsuut. This was a major settlement and gathering place: two years later, a whaling captain would report 270 Inuit there. Quayle promised to return to pick up the men, along with their anticipated mother lode of blubber and baleen, no later than July 1852. Twelve men including the first mate volunteered. Money was almost certainly a major factor in their decision. Whalers were paid on the lay system, meaning their wages were a percentage of their ship's net profits, so there was a strong incentive to return home with a full ship. After a month in Cumberland Sound, the volunteers were apparently willing to gamble that they would neither starve nor freeze to death.[11]

While these men must have been among the more adventurous whalers, they were fairly representative of mid-nineteenth-century crews. Eight of the twelve were single; most were in their early twenties. Ten were described as Americans, a relatively high percentage in the cosmopolitan whaling world.[12] They were all from the northeastern states. Most of the men were likely farm boys, factory hands, or unskilled laborers—apart from the first mate who came from a whaling family and had been at sea since the age of thirteen. The cook William Bandwell was the lone African American. He would have been among the lowest paid of the crew members, with the least opportunity for advancement.[13]

The twelve volunteers unloaded their personal belongings and ship's supplies, went ashore in two small whaleboats, and watched the *McLellan* sail away. *Aujaq* was fading; the days were getting shorter and the nights colder. Their first act was, literally, to build a home. Greenhand volunteer George Tyson described this in the vocabulary of American resettlement: the men "pre-empted a section of land whereupon to build a hut or house." They had taken some lumber from the ship, but it was not enough for a cabin. Instead, the men constructed a shelter out of the "stone of the country." They packed the walls with turf, laid poles across the top, and roofed the shelter either with canvas or with sealskins sewn together. A section of the roof was made "of the entrails of the whale," to allow light to pass through. This design seems to have owed much to Inuit *qammat,* which were dwellings with sod walls, framed with wood or whale bones, insulated with heather, and covered with skins that could include translucent sections of dried intestine. Inuit women presumably sewed the roof of the *McLellan* house; if not, it almost certainly leaked. After the snow came, the crew would bank walls six feet thick around their house.[14]

The Americans fared better that winter than many who followed them in the years and decades to come. Their captain had given them all the food he could spare, but it was not nearly enough to sustain twelve men for eight or nine months. Out of necessity, they traded for Inuit food and clothing, and they ate more nutritious food and lived more comfortably than on many whaling ships. They "occasionally secur[ed] a seal" for themselves, but were in large part dependent on Inuit, who shared game in exchange for Qallunaat goods. The Americans thrived on this diet; according to one of them, they "never were more healthy" and actually "increased in weight." No one contracted scurvy or any other serious illness. When their coal ran out, Inuit helped them collect whale skeletons from previous kills. The bones were easy to chop and

FIGURE 1.1. Taken at Qimiqsuut, this photo likely shows the remains of the *McLellan* crew's house. Photo by Philip Goldring, 1984.

apparently made a good fire. According to the American explorer Charles Francis Hall, burning whalebones produced a great heat, and he speculated that "one cord of bone must be equivalent to four cords of live oak." In 2008, Inuit Elder Pauloosie Veevee said he had heard that there is lots of oil inside bowhead ribs when they are fresh.[15]

When the bowhead whales returned in the spring, the *McLellan* overwintering party killed sixteen or seventeen of them. The Americans stripped off the blubber and baleen and left the rest, but Inuit ate parts of the whale and stored large quantities of meat in sealskin bags on various islands, likely for dog food. With perhaps $60,344 in oil and baleen put up—at a time when a single man could live in New York on roughly $250 a year—the Americans waited for their ship to pick them up.[16]

It never arrived. On its return voyage, the *McLellan* was smashed between ice floes "in terrible conflict" and wrecked in Davis Strait. The men waited through *upingaaq* and well into another *aujaq*. Having nearly exhausted their supplies, they struggled to find enough to eat, relying in large part on Inuit to keep them alive. They also ate whale meat from the bags that Inuit had cached. They heedlessly helped themselves to this old meat and were "very glad, indeed, to get it."[17]

Finally, in September 1852, a year after the men had been left behind, the *McLellan*'s captain arrived in Cumberland Sound on a British whaling ship, the

Truelove. He became alarmed when the *Truelove*'s crew spotted an American whaleboat full of Inuit, wearing American clothing and carrying rifles. But he need not have worried. Two days later they found the twelve volunteers, who were clad in Inuit clothing made of fur and skins. The men loaded themselves and their oil and bone onto the *Truelove*. At least from the perspective of a British whaler, "The Americans and natives seemed very loth to part, having been so long accustomed to each other's society.... I believe it would have taken some of our friends very little persuasion to stay another winter."[18]

What happened to the voyage's profits? The *Truelove*'s crew came out well: they agreed to transport the oil in exchange for being paid out for it at the same rate as their own. It's unclear what the *McLellan* volunteers received, as they were no longer official employees since their ship had been wrecked. I have found no evidence of any of them receiving a life-changing payout in 1852. The Inuit were paid only with "whale boats, belonging to the Americans, and some harpoons, lances and lines, rifles, ammunitions, etc."[19] These were items that Inuit truly wanted, but during commercial whaling low-ranking Qallunaat were not compensated fairly for their hard labor, and Inuit even less so.

Even though the Americans likely failed to receive their expected windfall, at least three of them returned to Cumberland Sound and built careers there. William Sterry was a reportedly cheerful man who would also join several more Arctic whaling voyages. He learned enough Inuktitut to get by, and at least twice left his ships voluntarily and took up residence in Inuit homes. We do not know what the Inuit families thought, but, ten years later, Sterry said he had "never enjoyed himself better—or had better health" than when he was with them. He added that he would "like a piece of raw seal meat" right now. Sidney Budington, the first mate, and George Tyson, a greenhand, both became renowned Arctic whaling captains. They went on to spend at least eleven winters each up north. We will encounter Budington and Tyson many times in these pages. Although their ties to Cumberland Sound were strong, they always remained ambiguous. Indeed, it is Budington's dismissal of the Arctic as "unfit for a white man or any one else to live in" which opens this book.[20]

The voyage had a ripple effect in Qallunaat whaling communities. The following summer, the *McLellan*'s owners outfitted two ships and sent them north to spend the winter. They were captained by Sidney Budington and his uncle James Monroe Buddington, and both ships returned full of oil and baleen.[21] British ships sailed that year as well. By the time Peter arrived eight years later, overwintering was commonplace. Most Qallunaat returned home safely, even though many of the factors that the *McLellan* men had in their favor—fresh Inuit food, fur clothing, relatively small numbers compared to the local population—were advantages that later whaling crews lacked. Still, Arctic whalers largely failed to enter American popular consciousness. Instead, the disappearance, suffering, and eventual death of the Franklin expedition dominated news of the North, and when whaling captains made headlines it was usually as Franklin searchers.[22] In 1860, would-be explorer Charles Francis Hall interviewed whalers in New London, Connecticut, before embarking on his first journey north. After speaking with several captains, Hall concluded, apparently with some relief, that "every Whale Capt . . . is confirmed in the opinion that White men can live with the Esquimaux—*It has been done time & again.*"[23] Hall was far more knowledgeable about the Arctic than most Americans; he was acquiring and reading every piece of Arctic literature he could find.[24] The fact that even Hall was not sure initially that outsiders could subsist in Inuit communities on South Baffin Island suggests that whaling stories failed to form much of a counterpoint to growing stereotypes about the Arctic's bleakness and hostility. Nor were whalers immune to this discourse. As *aujaq* drew to a close in Cumberland Sound, many inexperienced whalers also wondered if they would survive an Arctic winter.

Ukiaksaaq (Early Fall)

Ukiaksaaq sees the onset of snow, ice, and darkness. The sun begins to rise later and set earlier than in temperate regions. Freshwater ice skins the lakes, thaws, reappears, and gradually strengthens. Seawater becomes thick and slurry along the shorelines. Snow covers and uncovers the ground before finally settling in for the season. In 1860, this was the time of year when a group of American whalers would find themselves in dire circumstances and in need of help.

One month after Peter jumped ship and was dragged back onboard the *Antelope,* a group of American whalers made an unquestionably serious attempt at desertion in Cumberland Sound. On the evening of August 4, 1860, nine men from the ships *Ansel Gibbs* and *Daniel Webster* packed up their few belongings and rowed away in an open whaleboat. No one saw them go, even though it was still *aujaq* and there was no real night. The sun set around 10 p.m., rose around 3 a.m., and provided substantial twilight in between. Their escape was possible because the men from the *Ansel Gibbs* were the group keeping watch at the time and made their break while their captain was smoking cigars with a friend on a neighboring vessel. The deserters rowed under the bows of the *Daniel Webster* and the other two men dropped in. Within the hour, the captains sent out whaleboats to bring the runaways back. They never found them. The whaling ships were anchored among many small rocky islands. In

FIGURE 1.2. Sketch of lancing a whale in Cumberland Sound, showing the type of whaleboat in which John Sullivan made his escape. Timothy C. Packard, F 6870.3, Daniel B. Fearing Logbook Collection, Houghton Library, Harvard University.

the foggy twilight, the searchers could not make out a boat. Crews continued to comb the sound for a week, but the men had disappeared.[25]

According to the logbook keeper of the *Ansel Gibbs,* the two captains consoled themselves with the conviction that the deserters "will never live to get anywhere." This choice of words reflected the popular notion that the Arctic was not a part of the livable world, that it was an undifferentiated and unknowable space, that it was nowhere. The deserters' desire to get "somewhere" would cost three of them their lives.

One of the survivors, John Sullivan of South Hadley Falls, Massachusetts, later wrote up an account of their journey at the request of a magistrate. The understated eloquence of Sullivan's deposition, written in his own hand, suggests that he was better educated than most foremast hands and more literate than most of the officers keeping logbooks and journals. Sullivan testified that he had left home and tried to find "a berth to suit [him]" on a whaling ship, but as a greenhand with no experience he had little choice. He somewhat reluctantly signed on with the *Daniel Webster.* In Cumberland Sound, his crew anchored close to five other American ships.

On the night of their escape, Sullivan, his shipmate Warren Dutton, and the other men sailed more than fifty miles across the sound to the vicinity of Qimiqsuut. The deserters then hurried out into Davis Strait. Desertion was a problem on many American whaling voyages: from 1843 to 1862 the rates ran as high as 29 percent, though it was much rarer in Cumberland Sound. Any whaler caught deserting his ship could be punished by law or by unofficial methods such as flogging, which was illegal yet common.[26] Three days later, however, the deserters were hungry enough to approach an American whaling ship heading north. Its captain was Sidney Budington. When Budington first saw the boat with the black sail in the distance, he believed it was a whaleboat he had left behind the previous year, now captained by Inuit. His first mate disagreed, and the two men wagered tobacco and cigars on the question. Finally, the runaways pulled alongside and revealed themselves as Americans. Budington took them on board and fed them. John Sullivan believed Budington would have offered them jobs, but the runaways did not want to spend a winter in the Arctic. They wanted to get home.

The deserters were ill-equipped by any definition. They did not know this region. Only two of them had been to sea before. Sidney Budington gave them a chart along with some extra food and ammunition, but none of them knew how to navigate, and the only instrument they had was a small compass. One man who owned a quadrant and had experience as a navigator had been keen to join them, but as the captain's steward, he had been out visiting the other ship on the night of the escape and could not get away.

The men ate ravenously of the hardtack and salt beef that Budington's cook dished up. They had very few provisions of their own. Most whaleboats held basic tools like candles, a flint, a lantern, and a hatchet.[27] They had also taken a few blankets, and a portable oven called a "conjuror." John Sullivan and Warren Dutton had stuffed a small bag and a pair of wool undergarments with about twenty pounds of bread from their ship's hold. They had also stolen some cooked provisions, two guns, some harpoons and lances, and two tubs of whale line — but threw one tub overboard almost immediately to lighten their load.

The runaways fared quite well at first. They shot a small duck and a polar bear, although they were only able to drag the hindquarters of the bear into their boat and had to leave the rest behind. Following navigational advice they received on board Budington's ship, they successfully crossed Hudson Strait in rough weather. They were suffering by the time they reached the coast of Labrador. For the previous fortnight, their only food had been scanty rations of berries and mushrooms, which they had gathered ashore between arduous periods of rowing. They had encountered no more people.

A month after leaving Cumberland Sound, Warren Dutton died of starvation. Before nightfall, Samuel Fisher "took his knife and cut a piece of the thigh, and held it over the fire until it was cooked." The other men hesitated but all followed suit. After Dutton's body was consumed, Samuel Fisher and his cousin Joseph went after John Sullivan. They told him they "wanted some meat" and planned to kill him. Sullivan was carrying a small knife, and he managed to stab Samuel Fisher in the throat, knocking him over. He warily bent down and noted that Fisher was still alive. Sullivan began to cry. He "did not know what to do." The other survivors found him there, and together they waited a full day until Samuel Fisher finally passed. Then Joseph Fisher butchered his cousin and the men consumed the body.

The next day, the men abandoned their boat. Across the North American Arctic, most Inuit lived close to coastlines where they could reliably hunt marine mammals. But these Americans were used to farmlands and forests, to getting sustenance from the land, not the sea. They decided to walk overland. It took them four days to walk four miles, and then they saw water again. They were on an island. They crossed back and found their boat was wrecked. All they could do was, as John Sullivan put it, "remain there until we would die or be picked up." It began to rain, and then the rain turned to snow. For several weeks they had been "suffer[ing] very much from the cold"; now it was getting worse. *Ukiaksaaq* had arrived. Without transport or know-how, the men could not find anything to eat in this country and this season. Instead, they began eating their clothing: boots, belts, sheaths, and some bearskin and sealskin articles.

This environment may have appeared barren to John Sullivan, but there were local inhabitants. A passing boatload of Inuit picked up the party, and the men were eventually dispersed among various settlements. Missionaries reported that at least some of the runaways behaved ungratefully and "conducted themselves 'shamefully'" that winter. In July 1861, John Sullivan found a passage home to Massachusetts, around the same time he could have gone south with any number of ships leaving the whaling grounds.

John Sullivan's tragic story contains starvation, shivering cold, cannibalism, and a desperate march toward home and safety. These are all stereotypical elements of Qallunaat Arctic tales, yet I know of no whaler who experienced this level of horror in Cumberland Sound. Sullivan ran away because he was afraid he would die in the Arctic. When his story was retold, it likely provoked more fear of the Arctic. In other words, the idea that the Arctic was "unfit to live in" could be self-perpetuating.

Sullivan's story also demonstrates just how crucial it was to have local people to provide the resources and knowledge that new whalers lacked. Particularly in the colder months, visitors to the Arctic needed help.

Ukiaq (Late Fall)

As John Sullivan and the other deserters were waiting to die, their old shipmates were choosing their winter harbors in Cumberland Sound. *Ukiaq*, or

FIGURE 1.3. The small schooner *Franklin*, captained by Sidney Budington, at Naujaaqtalik (Niantilik) on October 1, 1863. This painting captures the feeling of isolation common to many whalers, but not the other people and ships that were likely nearby. © Mystic Seaport, #1939.1603.

late fall, is the season when the sea freezes up. At the beginning of the season, ice would form around whaleboats if they were launched. Men still saw whales but could not chase them.[28] They kept busy by sawing ice away from their ships before raising the anchors for the winter. This was "a cold and disagreeable job," wrote a journal keeper on the barque *Andrews* in 1865. "You will get your clothes wet everyday and of course you must suffer." When men had to work outside at this time of year, they kept an eye on each other's noses to make sure they were not frostbitten.[29] In the mid-nineteenth century, *ukiaq* seems to have begun in early to mid-October and lasted until November or early December, when whalers recorded walking across solid ice to visit neighboring ships.[30]

Ukiaq could be a period of scarcity for Inuit. Hunting was uncertain and risky when the ice would neither support a person nor allow for the passage of a kayak. This season could be treacherous for whaling crews as well. In 1866, Peter's old ship, the *Antelope*, was driven into the rocks during a snowstorm and had to be abandoned. Surviving records mention at least eight shipwrecks from 1859 to 1870.[31] How did whalers cope with the loss of their ships, which were supposed to be their homes for the duration of the voyage? And what were conditions like for those who could remain on board?

In 1867, the American bark *Andrews* was hammered up onto the shore by wind and ice at Nuvukallak, which the crew called Harrison Point. The men did "every thing that laid in our power to save [their] little home," but the ship lay fast on the rocks, and as the tide fell, the rudder forced its way through the casing into the hull. The *Andrews* listed on its side, "nearly on [its] beam ends." The ship was irreparable.[32]

The *Andrews*'s crew loaded bread and flour into whaleboats, trying to get at least the most basic supplies ashore. They evacuated by climbing down the hawsers or ropes that they had used to fasten their ship to land, then decided only three hours later to climb back on board. "There was none of us could say that it was very agreeable [on shore]," wrote one journal keeper, "for it blew heavy most of the time and the snow drifting." On the ship, the tide ebbed and flowed in and out of the men's berths. When one of the officers went down to check on the crew, he found probably over twenty men crowded into a dry space of about eight by eight feet.[33]

The crew of the *Andrews* spent over a week trying to sleep on the ship. They continued to unload coal, and cut away masts and other wood that could be salvaged. They carried goods to shore across an uneven rocky area covered in three feet of snow. They saved most of their supplies and belongings, but it was miserable work. Many had no dry clothes to put on, since everything they owned had been "well soaked with saltwater" on the first day and then frozen hard. When the ice became solid enough for travel, the shipwrecked crew moved three miles to Uummannarjuaq (Blacklead Island), where there were two "poor old" houses built by other whalers in previous seasons. The men moved into the houses and banked snow around them. They used wood from the *Andrews* to add berths to one building, and turned the other into a cookhouse and dining hall. Reflecting the attachment whalers often had to their ships, when the *Andrews* was abandoned, its logbook keeper ended the official record with, "So ends this voyage and a good home."

The *Andrews*'s men did not labor alone. Uummannarjuaq was also home to an Inuit settlement and three overwintering whaling ships. News of the shipwreck spread quickly. Four men from a Scottish brig came to offer their assistance, and the *Andrews*'s captain and an Inuk walked across the ice and returned with two more Inuit. Other Inuit helped the men sled their belongings over the ice to their new home, using some of the ship's sail canvas to make harnesses for their dogs. On Uummannarjuaq, Inuit built the men a snowhouse for storing their coal, and allowed several of the crew to live with them while the new berths were built. The captain of another American ship purchased the *Andrews*'s two-thousand-pound anchor, and the following March the first mate, two Inuuk men, and a team of twenty-three dogs moved it over fifty miles across the sound.[34] No deaths were recorded among the *Andrews*'s crew that winter, and they were later joined on the island by the crew of the

Isabella, who lived nearby in houses made of sails and snow after their ship ran aground.[35]

At the height of the whaling period in the 1860s, no one had to worry about being truly marooned—they could always get a ride home on another ship. No captain seems to have turned down additional passengers, although extra men limited the amount of whale oil a ship could carry.[36] Whalers ran a serious risk of losing their ships during freeze-up, but the communities that connected and surrounded them mostly kept the notorious Arctic calamities of starvation and death at bay.[37]

Although a ship could offer comfort and security, it was not comfortable. For those whose ships remained intact, accommodations on board were only slightly more luxurious than in the *Andrews*'s poorly constructed hut. Men did what they could to insulate their ships from the cold. They fastened sails over the decks like tents, or brought special cloth and lumber for that purpose.[38] They banked snow around the hull, both for warmth and to protect it from ice.[39] All ships had at least one coal stove on board, but its warmth did not spread through all the living areas. One captain had a stove installed in the crew's quarters during the second *ukiaq*, and the journal keeper commented, "poor fellows they have been wanting one for some time back."[40] Even George Tyson, one of the *McLellan* overwinterers who returned several times to Cumberland Sound, was unable to keep his own stateroom entirely comfortable on one of his last voyages in 1877–78. When he climbed into his berth in the colder months, he frequently found his blankets frozen to his bed. During a spring rainstorm, the cabin dripped with water that found its way through the wood that had contracted in cold temperatures.[41]

The windowless cabin below the ship's prow, known as the forecastle, was where lower-ranking whalers ate, slept, and socialized. Forecastles were anywhere between sixteen and twenty-five feet long, and many had ceilings so low that only short crewmen could stand up in them. Most of the space was taken up by the whalers' sea chests of personal belongings and by double-decker rows of narrow berths built into the walls. In the early twentieth century, a Canadian doctor described the forecastle of an Arctic whaling ship as "the most stifling place I was ever in," and thick with bedbugs. The forecastle was dark even during the constant summer daylight. Some forecastles had prisms in the ceiling, but the main source of natural light was a single hatch above the stairway, which was open only when the weather was good and presumably rarely in the winter. As the roof hatch was also the room's source of ventilation, the air must have been rank through the colder months.[42]

The only other light in the forecastle came from malodorous whale oil lamps, which released a thick, oily, black smoke; the stench of whale oil was legendary. One man recalled that even years after a whaling ship had been decommissioned, it still reeked with "that whale oil stink."[43] Some Inuit on northern Baffin Island considered whale blubber too repulsive to use in their lamps except in emergencies. Cumberland Sound Inuit mostly tolerated it and adjusted to the smell, as the whalers presumably also did, since they lived with it day in and day out.[44] Compounding the stench and smoke, many foremast hands washed only irregularly. This was unsurprising aboard an overwintering whaling ship, given that nineteenth-century soap did not lather in saltwater, and fresh water had to be chopped as ice, sledded to the ship, and then melted.[45] Conditions in the forecastle revealed that any complaints about dirt, close quarters, and lack of privacy in Inuit dwellings were ethnocentric and hypocritical.[46]

Whaling ships might not have been comfortable, but they kept ill-equipped men from freezing to death. As unpleasant as the forecastle could be, for many men, spending time outside was worse. Few whalers arrived in Cumberland Sound with adequate clothing. A first mate described some of the men on his ship as "shaking shivering nearly froze[n]."[47] Frostbite—especially mild cases that did not cause serious damage—seems to have been commonplace.[48] The deserter John Sullivan had feared that he was "very badly fitted out for such a cold climate," and he probably was.[49] Agents and outfitters often sold greenhands substandard goods at inflated prices, and most first-time whalers had little money to purchase clothing beyond the bare minimum. Men without good clothing could hardly be expected to find the colder months anything but forbidding or to think of Cumberland Sound as hospitable. Whalers often focused on earning money to make a better life back home, but their financial situation also affected their perceptions of Cumberland Sound.

Some men, particularly those who overwintered multiple times, were able to trade with Inuit for fur clothing. If everyone had done this, it is unlikely that Inuit could have met the demand, given the number of caribou skins needed and the labor involved.[50] For those lucky enough to obtain Inuit clothing, it was adequate protection in almost any weather. The anthropologist Franz Boas, clad in caribou-skin clothing sewn by Inuit women, experienced one day of −54°F (−48°C) in Cumberland Sound in 1884, but commented that he "easily tolerated" this temperature.[51] Captain George Tyson forbade his crew members from going out for extended periods until they obtained good-quality, presumably Inuit-made, winter clothing.[52]

Even whalers who rarely left the ship would not have been isolated from the world outside. They frequently encountered Inuit and also sought them out. With motives that could have ranged from friendship to lust to obtaining furs, in 1867 the men on the *Isabella* stole bread out of the hold during the night, either "selling it or giving it away by the Bagful" to Inuit who came from a place the whalers called Molly Kater-nuna, probably near the mouth of Nettilling Fiord.[53] Once the sea ice formed, Inuit came and went "continually" from the ships, and sometimes they slept on board.[54] A journal from the Scottish ship *Emma* reported as many as fifty Inuit on board.[55] Women were frequent visitors. When Inuit men were not whaling for the ships, they traveled and hunted even more intensively than before, offering or trading surplus game to the visitors. Women and children remained near the whaling ships to mend, sew, interpret, and prepare skins for Qallunaat.[56] The crews, who had little to do until the return of the whales the following spring, were grateful for the companionship, help, and diversion the women provided.

Did these Inuit women make Qallunaat whalers feel more at home? For some yes, for others no. Women were strongly associated with the idea of home for Victorian Qallunaat, as they are in most societies. Whalers often expressed pain at leaving their wives and mothers. For the youngest crewmembers, this was often their first time away from home. Parting never became easy. George Tyson wrote to his wife on one of his later voyages to Cumberland Sound, "I left you without saying Good Bye—but you know me and will excuse it—I cannot bear to say Good Bye."[57] Another whaling captain noted that he built an extra-large woodshed and chopped three winters' worth of wood for his wife before leaving each time, in case he did not come back.[58] Men worried about their family members and missed them, in some cases it seems with a near-constant ache.

However, while women were central to home, some Qallunaat drew a sharp distinction between Inuit women and what they called "civilized" women. Ambrose Bates, a first mate, judged men who took up with Inuit women as having "lost all respect for sisters Mothers wives and even themselves." For Bates, at least in his writings, Inuit women only served to remind him that he was not at home, and not with his wife. Yet others disagreed. Many of Bates's fellow whalers, he ranted, "seem to look upon [Inuit] as a superior race . . . and seek to lavish upon them the greatest favors possible." Bates met one Qallunaaq officer who forfeited his earnings to stay in Cumberland Sound. When his ship was full of oil and ready to depart, he chose to stay behind with an Inuk woman and forgo his large share of the profits. He worked for the next several years as a contract laborer for whaling ships for little pay. While most American whalers were highly motivated by profit and eagerly anticipated their return home, this man opted to remain in Cumberland Sound. He seems to have made a home there with his new family, but it did not last. In 1868, he returned to the United States alone. Unfortunately, we do not know why he decided to leave, or anything about the Inuk woman and their possible children.[59]

Countless Qallunaat whalers partnered with Inuit women, even if few men recorded it or spoke about it publicly. Some relationships, at least in a later period, were strong ones that continue to be remembered positively by Inuit. But the arrival of hundreds of Qallunaat men upended lives. Several accounts survive of violent altercations over Inuit women. Some of the most damning are in an account by Adam Bek, a Greenlander who lived in Cumberland Sound from 1861 to 1863. Bek was an Inuk. He identified as a Christian and contrasted himself positively with the non-Christian Inuit in Cumberland Sound. Bek writes that after an American whaler moved into a tent with an Inuk woman, her jealous husband and his friends cut the tent with a knife, stole the American's extra clothes and possessions, and drove the pair out into the winter night. Their eventual fate is unknown. In another case, a woman, Kaunaq, moved in with a whaler named Karl on board his ship. After five days, Kaunaq's husband came to get her back. The two men both grabbed onto Kaunaq and pulled her. During the tussle, the husband drew a knife, and Karl hit him over the head with an axe. The captain stepped in and ordered Kaunaq to leave with her husband. The people on the ship watched the Inuit couple retreat to their dog team. Then they saw the husband slash Kaunaq in the face with his knife. "Their tracks were very bloody from her wound in the head," reported Bek, "as if someone had flensed a harp seal."[60]

It is impossible to tell how many rapes and assaults occurred in the early years of whaling, although we can assume that Qallunaat whalers committed many violent acts that they chose not to record in their usually sparse records. Very little is known about the ways individual Inuit men perceived cross-cultural sexual encounters in this early period, whether fierce jealousy was a common reaction or just the rare reaction that Qallunaat noticed. I know even less about the women's motivations, or the extent to which such relationships were forced or consensual, or how the initial arrival of so many Qallunaat affected Inuit feelings about family and home.

Ukiuq (Winter)

After freeze-up, *ukiuq* set in. This was true winter, very cold and dark. Explorer Bernhard Hantzsch later described *ukiuq* in Cumberland Sound as a time "when all land and sea are hard-frozen, and the wildlife which has not migrated is in its full winter dress, when the seals no longer come out of the water, and the caribou antlers are ossified."[61] Temperatures varied, but days below −20°F (−29°C) were common. The sun did not come up at all throughout most of December and early January, although the *ukiuq* sky was often streaked with undulating tails of northern lights. It was probably the season that whalers feared the most, although it is hard to be sure since most logbook keepers considered "winter" in Cumberland Sound to extend for at least eight months of the year. Even during the warmer months of constant daylight, one journal keeper wrote poetry describing a "barren snow clad land," where the winds "blew drear and stark."[62] Whalers saw themselves as braving the "desolate shores of a frozen zone" and venturing into "icy solitudes of the Arctic, far isolated from our homes and friends."[63]

Their anxiety was somewhat justified: *ukiuq* was cold and icy, and Arctic whaling was a dangerous job. There is at least one recorded death in every surviving American Cumberland Sound logbook from the height of the industry in the 1860s. The average was two deaths per ship, approximately one in ten to fifteen, depending on the size of the crew.[64] Dozens of men also suffered from nonfatal illnesses or injuries incurred from accidents, fights, or, in one case, being run over by a sled load of meat.[65] But most whalers survived the winter. *Ukiuq* was, statistically speaking, the safest season to be on a ship in Cumberland Sound. It was not unheard of for Arctic whalers to venture away from their ships and freeze to death, but in all the extant American logbooks and journals from Cumberland Sound, there is not a single employee death recorded between the months of November and March.[66] It was impossible to drown at this time of year. The ice was solid and safe. No one could be knocked overboard or caught in the line while chasing a whale. Whaling ships carried enough provisions to last the winter, and if the captains decided to stay on another year, they could often trade for what they needed from other ships. Few men on American ships were recorded as sick during *ukiuq*, perhaps because bacteria could not multiply as effectively in the cold air. Most whalers took few risks in wintertime. Unlike Arctic explorers and scientists, they were not concerned with mapping, sledging, or carrying out scientific observations in subzero temperatures. Captain George Tyson wrote of winter on a frozen-in ship in the Arctic regions: "One at home cannot imagine how dull this life is."[67]

Still, at least some whalers reserved an abhorrence for Arctic winters. One of the most vehement examples is the 1867–68 journal of Ambrose Bates, the first mate on the *Milwood* who so disapproved of relationships with Inuit women. Bates, then in his mid-thirties, had a typical background for a career whaler. He had grown up in a farming family before going to sea at the age of nineteen. He mixed long periods of life at sea with varied employment in the United States, including as an undertaker and a butcher. He overwintered at least twice in Cumberland Sound and once among Inuit in Hudson Bay. Bates never learned to see the Arctic as home. His journals consist largely of letters and poems to his family, in which he complains about his surroundings and longs for New England. His writings evoke a pervasive sense of discontent and unsettledness. One of his verses to his wife repeats a recurring theme:

> When I upon the ocean roam, I long for thee;
> When blessed with all which makes a home, I miss the sea.[68]

Bates once described his ideal home as a ship, but one that he would captain, and where all his loved ones—his "little home circle"—would sail with him. He would never realize this dream.[69]

As a first mate, Ambrose Bates did not have to write his journal in a shared berth in the forecastle. He had his own small stateroom, which was probably no larger than six feet by four feet but which accorded him privacy—a true luxury on board a whaling ship. Unlike the windowless forecastle, it had a sort of skylight in the roof and a porthole on its outside wall. On his previous voyage to Cumberland Sound, Bates had described his stateroom as a domestic, personal space. On one wall hung a small mirror, a case containing a brush and comb set, and some articles of clothing, including slippers. A berth was built into another wall, and over it Bates stored his rifle and pistols. The room was too small for a chair. Bates composed his poetry while seated on the sea chest that presumably contained, among other things, mementos of his wife and son. Men often went through the keepsakes in their sea chests once a week, on Sundays.[70] In spring 1868, Bates decided that he could no longer live on board the ship, due to an unspecified disagreement with the captain and possibly the other officers. Interpersonal relationships and hierarchies sharply affected whalers' ability to feel at home on a whaling ship. Bates moved into a house on Uummannarjuaq (Blacklead Island), probably one of the buildings that had housed the *Andrews*'s crew after their shipwreck.

Bates and the other descriptive American record keepers in Cumberland Sound had much to say about the Arctic landscape. Most of it was unflattering. Whalers appreciated one local resource: whales. But they judged the region's capacity for subsistence by northeastern American standards and found it lacking. Ambrose Bates could see only that there was no farmland, no forests; the soil was "scanty" and produced only "weeds and sorrels."[71] "What a country this is," he wrote in 1868, "where rocks will not burn and there is no wood nor coal or anything else that I know of."[72] This is a particularly oblivious statement, considering that Bates had been sent to Cumberland Sound to accumulate whale blubber, which Americans burned and which was probably at that very moment fueling a lamp in his stateroom. He was probably also aware that Inuit burned seal blubber for heat, light, and cooking purposes. While

Bates and most career whalemen spent only a fraction of their adult lives in the United States, it could still be difficult for them to see an Arctic climate as potentially hospitable, as providing the same necessities albeit in different ways.

What the whalers saw, thought, and wrote must also have been limited by the discourse of a barren, hostile Arctic to which their generation was becoming accustomed. Bleak and horrifying descriptions can actually be comforting in their familiarity, as they make order of a new landscape and reaffirm one's place in the world.[73] Whalers may have embraced the existing language of desolation partly because it was such a good reflection of their own homesickness. To sailors whose loved ones could only be reached by water, being frozen in for the duration of *ukiuq* was both a literal and symbolic marker of isolation. Ice cut them off. In the United States, the newly expanded postal system meant that most Americans enjoyed regular correspondence with absent friends and relatives.[74] Bates lamented that his letters could not travel over the ice to reach his family. "The ice field holds me fast," he wrote.[75]

Ice had very different connotations for Inuit: freeze-up was a time of renewed connections. It was easier and faster for Inuit to travel over ice than water, so much so that some Inuit refer to November as *tusaqtuut,* or "a means to hear." This was the time that dog teams set out, when "people would hear from each other after the ice had formed." A contemporary Inuktitut dictionary defines winter in part as a time when "one can travel, go in any direction."[76]

Bates, too, learned to see the ice as a mode of transportation: he journeyed up and down the sound by dogsled, paid visits to other settlements and ships, and broke up the monotony of the winter months. But as he expressed it, coming back to his ship was "not getting home." His travels only served to "lessen the great space of time which divided me from my real home," which was, he wrote, wherever his wife was. He measured his time remaining in Cumberland Sound not with Arctic seasonal markers but with New England ones. He would see his wife again "when twice the maple trees have been clothed in red. And twice in green . . . And twice the harvest shall be gathered to the store-houses."[77] Other whalers also spoke of missing specific seasonal events, such as certain flowers, the ripening of corn, and the fall harvest.[78] And even though many New Englanders found their own winters desolate, for Bates they had become a symbol of warmth and family. In Cumberland Sound, thought Bates, the cold was almost unbearable. In New England, it was "sweet to lounge on winter's night / By the hearth stone blazing bright."[79] For Bates, the physical landscape was indissoluble from the landscape of the mind; each flowed into the other.

Most whalers' descriptions of the Arctic only skimmed its surfaces. It would be unfair to expect them to appreciate such an unfamiliar place as Cumberland Sound when they inevitably lacked both the words to describe it and the knowledge to understand it. Yet the region appeared neither empty nor desolate to local Inuit; it was home. Physical and mental landscapes also converged for Inuit, but in different ways. The land, sea, ice, animals, and people of this region were, and continue to be, enduring repositories of history. To use the Arctic anthropologist Mark Nuttall's term, the region was and is a "memoryscape" to its people, a physical expression of local culture.[80] Inuit themselves were part of a larger network of people's names passed down through generations. Inuit stories—most of which were unknown to the whalers—filled up the expanses of rock and water and ice. These stories made the landscape familiar, tied people to place, and created a homeland.[81] Inuit also used place names as both navigational aids and landmarks for stories. The anthropologist Franz Boas documented 930 place names in Cumberland Sound in the 1880s; local Inuit were able to confirm the majority of these names a century later. Today the organization Inuit Heritage Trust has collected over 9,000 place names across the territory of Nunavut, with hundreds in Cumberland Sound.[82] Qallunaat whalers adopted corruptions of a few of these names, but they also bestowed their own place names, often in honor of captains or ships.[83] Naming helped whalers situate themselves and made Cumberland Sound more familiar, but their names never approached anywhere near the depth of Inuit toponymy. And finally, while some whalers wrote of this region as a godforsaken place, Inuit in this period perceived hundreds of supernatural beings known as *tuurngait,* the vast majority of which were helpful and benevolent. *Tuurngait* were individually identifiable, with names, homes, and specific spheres of influence. They were everywhere.[84] In the face of this abundance, the formulaic icy descriptions of whalers seem particularly void and cheerless.

Even many career whalers, at least at times, described Cumberland Sound as another planet. One whaler felt that he was "out of the land of the living."[85] Captain George Tyson, after choosing to spend more than a decade of winters in the Arctic, wrote that whalers were "isolated from all the world" and "must make a world of [their] own." Tyson meant an imaginary world: he described his own mental picture, with its temperate "grassy lawn of the newborn spring" and "budding of the trees."[86] But whalers did not just live in their thoughts; they also transplanted their tangible, cultural world into the existing Inuit one. Men read and played checkers, cards, and backgammon. Various crews hosted dances on board ship, held a "Grande Masquerade Ball," and

FIGURE 1.4. Songs and dances learned from whalers remain part of Inuit culture today. Elisapee Ishulutaq, *Children Dancing to the Whaler's Jig*, 1983, stencil on paper. Collection of the Winnipeg Art Gallery, Gift of Indian and Northern Affairs Canada, G-89-1374. Printmaker Josea Maniapik, photographer Leif Norman.

built a stage where they performed plays in front of as many as two hundred people. In the 1870s, one journal describes a litany of dances, recitals, minstrel shows, prayer meetings, fencing, plays, lectures, and debating clubs. By this time more captains were insisting on fur clothing, and the crews constructed an outdoor gymnasium complete with ropes, rings, and dumbbells. Some played baseball, went skating, and played tag.[87] Christmas dinner on board the *Milwood* in 1867 would not have seemed out of place in New England: it consisted of clam chowder, boiled ham, roast pork, boiled potatoes and onions, gravy, cranberry sauce, plum pudding with sugar sauce, plum cake, and apple and cranberry pies.[88] These activities were part of creating a sense of home. They also changed Inuit homes, as families learned songs, words, dances, games, recipes, and other traditions that remain integrated into local culture today.

Whalers also learned from Inuit. They sprinkled Inuktitut words throughout their journals. They tobogganed on sealskins. At least one traditional Christmas dinner menu was augmented with "two splendid seals, roasted whole and stuffed," each with a potato in its mouth.[89] In making "a world of their own," the whalers were making a new world with Inuit. Within a few decades, this world would be almost entirely without whales, and the reason for its creation would be gone. For Inuit, Cumberland Sound would remain home. For nearly all Qallunaat whalers, it was a temporary home at best, and one usually suffused with absence and longing and a plan for leaving it.

Upingaksaaq (Early Spring)

Upingaksaaq, or early spring, is a beautiful time of year in Cumberland Sound. Captain George Tyson did not think that "any climate in the world" was better than the Arctic during *upingaksaaq* and the season that followed. The days became longer than they do in the United States. The sun rose high in the sky with a "peculiar silvery whiteness, like a burnished silver mirror."[90] The snow began to melt, and eventually, pools of water the color of sky appeared on the surface of the sea ice.[91] The soft hum of running water returned, running through the background of young seal hunting and preparations for spring whaling.

As water became easier to obtain and temperatures rose above freezing, the captains ordered their crews to scour off the grime that had accumulated over the winter months. This could be a substantial task. In May 1861, an entire crew spent three days "cleaning up the filth about the ship" and "carving off the dirt from around the ship."[92] On some vessels, however, men were unfit for this work: they were suffering from scurvy. The deserter John Sullivan had heard that "so many men" had died of scurvy in Cumberland Sound the previous winter, and he and his fellow runaways were "afraid to remain there, for fear that we might get it."[93] Survival rates were higher on Sullivan's whaling ship than in his band of deserters, but four of his shipmates would indeed die of scurvy, and many more were debilitated with it for months at a time. Scurvy was a concrete consequence of Americans' failure to understand Cumberland Sound as a hospitable place. Many whalers suffered and died from an overreliance on shipboard supplies, even when healthier local foods were available.

Scurvy results from a long-term deficiency of ascorbic acid. Even the most severely afflicted scurvy patients make quick, full recoveries if they are given anything containing a good amount of vitamin C. In the Arctic, the most readily available source is raw or lightly cooked meat. Favored Inuit foods like whale skin and raw seal liver are particularly high in it. The monotonous whalers' diet—salt pork, salt beef, bread, flour, and coffee—failed to provide enough vitamin C. The additional "small stores" recorded by one ship were not a strong source either; they included butter, cornmeal, dried apples, molasses, beans, sugar, oatmeal, and rice.[94] Reflecting the tedium of the usual diet, one logbook keeper noted that there was pie for supper one night in June. He underlined his happiness by again writing PIE in bold letters at the end of the entry.⁹⁵

John Sullivan had been wrong about one thing: men did not usually die of scurvy in the winter but rather in *upingaksaaq*. It took time for whalers' bodies to use up their stores of vitamin C. Initial symptoms of scurvy—diarrhea, weakness, and fever—did not appear until at least three months after the last intake. Early sufferers also complained of irritability, foul breath, and loss of appetite; but these complaints would have been unextraordinary for men eating similar food every day and living in a filthy, cramped forecastle. The American logbook keepers typically began to record instances of scurvy when its more serious symptoms appeared after five to seven months, between February and May. Signs included sore joints, purple blotches on papery skin, bleeding swollen gums, loose teeth, the reopening of old wounds, and severe pains and paralysis in the legs. After eight to nine months without ascorbic acid, victims were susceptible to fatal cardiac hemorrhages.⁹⁶

Scurvy must have been emotionally as well as physically devastating for men who lived and died in such close quarters, watching each other becoming weaker, not knowing whom the disease would strike next. On board John Sullivan's ship the *Daniel Webster* in 1861, the first mate, Charles Frasier, was showing signs of scurvy by mid-February. A week later, he had "a lame leg." By mid-April, there were seven men on the sick roll, and on April 24, Frasier died. The men who were fit enough made a coffin and climbed the hill behind their harbor on Qikiqtat (Kekerten Island) to dig a grave in the frozen ground. The next morning, they took Frasier's body ashore and buried him in the earth. Two days later, the desperate captain set out across the sound to "try to get some seals for the men to eat." By now, six of the men were "lame," and eight had "some scurvy in gums," meaning that their gums were rotting and spongy and perhaps swollen enough to completely cover their loosened teeth.⁹⁷ The captain returned with eight seals, and there were no more deaths. By the middle of May, all were recovering well.⁹⁸ Yet the following year, again in *upingaksaaq,* scurvy returned to the ship. This time it claimed three lives: the second mate Emil Bessuell, the boatsteerer George L. Wiser, and the steward Samuel Watson. From the time they were first noted as being "quite lame" or "down with Scurvy," it took approximately two months for each man to die. They were buried next to each other, one by one, on Brown's Island, just across the mouth of Kingnait Fiord from their shipmate Charles Frasier.

These deaths were almost certainly unnecessary. The *Daniel Webster*'s logbook contains far fewer mentions of trade with Inuit than do other ships' records in the same years. It is possible that the logbook keeper did not consider these exchanges worth noting, but he recorded only two instances of acquiring seal meat during more than two years spent in Cumberland Sound. Meat seems to have been available for other ships.[99] During February, March, and late May of 1861, the *Black Eagle*'s Inuit boat crew caught as many as eight seals a day; later they also brought caribou.[100] The captain of the *Black Eagle* had already overwintered elsewhere in the Arctic, and had perhaps learned about the value of local provisions on this voyage—other experienced captains certainly did.[101] None of the *Black Eagle*'s crew seem to have suffered from scurvy, even as men were dying on the *Daniel Webster* and other ships in the sound.[102]

Why were these deaths allowed to occur? First, confusion about the prevention of scurvy persisted in this period.[103] Any variance between life on land and at sea could be a suspected cause. As late as 1911, a naval surgeon on Robert Falcon Scott's first Antarctic expedition declared that "tainted food" was the main cause of scurvy, followed by other conditions like "damp, cold, over-exertion, bad air, [and] bad light."[104] It seems that American whaling ships did not routinely administer lemon or lime juice, as the British navy did after 1795. In Cumberland Sound, first mate Ambrose Bates believed that the best way to remain healthy in the Arctic was to exercise, stay warm and dry, remain cheerful, and avoid becoming constipated.[105] Similarly, the men on the whaling ship *Ansel Gibbs* were ordered to run around outside every day, or walk between decks if they were too weak to leave the ship. They also went tobogganing, but this was not for personal enjoyment; rather, it was "sleighing down hill for the Scurvy." When May came and some men were still crippled with the disease, the captain ordered them put on a sledge and taken ashore. There, the swollen parts of their legs were covered in earth. The mate concluded, "We think it helped them."

Within two weeks, the men of the *Ansel Gibbs* were all healthy again. This presumably had more to do with something else recorded in the logbook: during those weeks and for the first time in months, Inuit began bringing large quantities of fresh meat to the ship.[106] The same mate who proposed the earth cure noted earlier that year that his "men that has got the Scurvy is some better" after he fed them "the blood and meat that came off the seal" that Inuit had brought to the ship.[107]

Given that fresh meat was at least widely suspected as a cure, why did officers not procure it for their crews throughout the winter? W. Gillies Ross correctly notes that this was not always possible: "There were countless reasons why hunting might fail."[108] It would hardly have been feasible for every ship in the busiest years to obtain a steady supply of fresh meat. Still, the prevalence of scurvy on Arctic whaling ships remains hard to understand. It seems to be one of those situations—which also exist in the present—when people work hard to justify actions that go against clear evidence and may even be against their own best interest. Desire to maximize profits was likely a major factor on some ships. Victorian industries are not known for their focus on the health and safety of laborers, and whaling was no exception. Some captains brought a few live pigs or some unsalted meat, but fresh meat was never a dietary staple.[109] Officers shot the odd eider duck or Arctic hare, familiar animals whose temperate counterparts they may have hunted in New England. They also killed polar bears for their fur, and sometimes for food.[110] The foremast hands could—in the right time and place and often with Inuit advice—return to the ship with "a mess" of Arctic char, clams, berries, or eggs.[111] But although whalers could dispatch sixty-foot marine mammals from twenty-eight-foot boats, they were not reliable hunters of anything else.[112] The only real option for obtaining fresh meat was to hire a crew of Inuit and their families on arrival, and keep them on through the winter. This added to the cost of the voyage.

Whaling crews had ready access to another decent local source of ascorbic acid. Before freeze-up, most ships had taken blubber on board. American whalers processed their blubber immediately, so they would have had to preserve some of it in its raw state to take advantage of its vitamin C. Most Scottish ships did not boil their blubber down into oil, and at least one crew ate it months later, with no apparent ill effects. One Scottish captain deliberately saved narwhal skin and blubber for scurvy prevention.[113] However, most Qallunaat were trained to see blubber as money rather than as food or medicine. A dead whale was something that would provide the funds to buy food later, not serve as food now.

At least three of the *McLellan* overwinterers later spoke up about the importance of fresh meat. Over the course of their voyages, they learned to see country food as healthy and made obtaining it a priority even if it cut into their profits. George Tyson hired enough hunters to supply his crews with seal meat. In 1877–78, this meant that his financial backers had to fund weekly American rations for up to forty Inuit men, women, and children—and his crew was about half the size of those on many whaling ships. He stated that his men had not suffered from scurvy that year because they ate as much seal as the Inuit. Many gained weight: the cabin boy went from 116 to 160 pounds, and the second officer could "scarcely see out of his eyes for fat."[114]

William Sterry, who shared Tyson's approval of the Inuit diet but not his rank of captain, was forced to take matters into his own hands. He left his ship when he felt himself weakening. For two months in the spring of 1855, he moved into a small camp with three Inuit families. Here, thirty miles away from his ship, he quickly recovered and was healthy. Sterry began a relationship with an Inuk woman and declared himself so content that he would not have returned to his ship at all if his captain had not come to fetch him when the ice broke up. It is impossible to know if the Inuit family was equally content—perhaps Sterry was a welcome guest, but when Qallunaat rely on Inuit resources and knowledge, it is never a neutral event. In any case, Sterry could not remain with the Inuit family: his labor was required on board, since fourteen of his shipmates had died from scurvy that year.[115]

Sidney Budington, another *McLellan* overwinterer and the captain of Sterry's ill-fated 1855 voyage, also learned from his experience. "I am not afraid of losing any more men while I have command over them by Scurvy," he reportedly said in 1860. "Whenever there are appearances of it aboard, I will have every Pork & Beef barrel—salt provision of every kind—headed up at once & every man shall live upon bread & *fresh* provision such as whale, walrus, Seal, deer, bear, ptarmigan, duck etc. etc." His edict was put to the test in 1861, when two of his crew members came down with scurvy. He arranged to send them out to live in an Inuit community. This could have saved their lives, but the sense of home that the ship could provide, with its familiar company and foods, was too powerful. One of these Qallunaak, an eighteen-year-old Frenchman incongruously known as John Brown, decided to return alone to his ship against the advice of his shipmate and two local men. Brown said he wanted to get back in time for Sunday dinner, because that was the one day a week the men got "duff," a pudding made with dried apples, unfortunately a poor source of vitamin C. He became lost on his way back and froze to death. Ideas of home intertwined with risk in complex ways for Qallunaat whalers: here John Brown was sent away from the ship to cure his scurvy, but his apparent overwhelming desire for the diet which had caused his scurvy resulted in his death.[116]

Scurvy would continue to plague American whaling ships in Cumberland Sound at least into the 1880s, at a time when some poor communities in the United States also suffered from it.[117] Since Inuit continued to eat fresh meat alongside whaling ship provisions, they presumably remained immune to scurvy throughout the whaling period. Unfortunately, the same could not be said for contagious diseases. At Qikiqtat today, the graves of Charles Frasier and the other whalers are not alone. The bodies of Inuit men, women, and children also lie under stones and pieces of ships' casks. Inuit lost children, brothers, mothers, sisters, grandparents. We don't know how many Inuit in Cumberland Sound died during the height of whaling, but disease caused devastating losses: probably at least one third of the population, and possibly more than half. Diphtheria, cholera, syphilis, measles, and various pulmonary afflictions swept again and again throughout Inuit communities after the arrival of whaling ships. Food shortages at certain times and places—caused in part by the increased pressure on local resources as both Inuit and Qallunaat moved into the region—would have exacerbated the effects of disease. Very sick adults were presumably unable to hunt or sew, further weakening their families.[118] The arrival of Qallunaat brought grief and loss into many Inuit homes.

Upingaaq (Spring)

In *upingaksaaq,* the American whalers finally prepared for the main bowhead whaling season. They had signed on and endured an Arctic winter for this. For impoverished whalers, or for those who felt they were not rich enough, a whaling voyage was a chance to make a better life for themselves and their families. Some hoped to purchase a farm or house or business in the area where their ancestors had lived. Most would return with relatively little. Nearly all unskilled labor in New England paid better than whaling; one study concluded that crewmen's earnings averaged only 20 cents per day. But unlike many American workers, whalers were employed full-time year-round, they received food and lodgings of a sort, and they could always hope for an unusually successful voyage.[119] Their future home life depended on the number of whales they caught.

The whale hunt was in full force by *upingaaq.* As the ice broke into floes and bumped and jammed its way out of the sound, it crossed paths with the bowhead whales streaming in from their early spring feeding grounds. Up to about sixty feet long and 220,000 pounds, *Balaena mysticetus* live in northern waters year-round and have the longest baleen and thickest blubber of any whale hunted in the nineteenth century.[120] While the whaling ships were still frozen into their harbors, men clambered out onto the ice. Inuit and their dog teams helped sled whaleboats, sailcloth tents, and provisions to outlying islands or to the floe edge, the place where the land-fast ice met open springtime waters.[121] In the days of commercial whaling, this floe edge ran approximately between what became the two major whaling stations of Qikiqtat (Kekerten Islands) and Uummannarjuaq (Blacklead Island).[122] From the beginning, not all the whales being shipped out were caught by Qallunaat whalers. Most ships also had at least one Inuit crew working for them by the 1860s. One Inuk, Tessuin, fielded his own boats and offered his catches to the highest bidder.[123]

The killing and butchering of whales by hand was intense and prolonged. George Tyson of the *McLellan* described a difficult hunt on his second Arctic voyage in 1855. The chase began as it usually did: the men spotted a whale, approached it in their open boat, and harpooned it. Their harpoon was not designed to kill the whale. It was attached to a line that joined whale and boat together. The injured whale would generally attempt to dive or flee, but would remain tied to the boat until exhausted and weakened. When the whale eventually faltered, the head of the boat crew would attempt to lance its vital organs and kill it. As soon as Tyson harpooned this whale, she dove and stayed down for an hour. When she resurfaced, "she began to beat the water with her flukes, and swirled around." Eventually, she broke the line and swam away, perhaps as fast as twelve miles per hour, the harpoon still protruding from her side.[124] Unwilling to give up, Tyson and his crew rowed after her. The moon rose, but in *upingaaq*, the night and day were hardly distinguishable, and it was possible to keep the injured whale in sight. It took nearly twenty-four hours for her to tire enough for them to approach her. Tyson struck a second time, recalling that "the water all around [was] covered with blood, and we knew she was done for. Three or four lances were hurled into her ponderous bulk, and at last our exertions were rewarded by seeing her roll over on her side. She was dead."[125]

The men attached another line to the dead whale, and towed her to the edge of the pack ice. The chase had taken them nine miles away from their frozen-in ship. They did not have the means to transport thousands of pounds of blubber back from the floe edge. Other Arctic whalers regularly dragged their butchered catch across similar distances by sled, or towed it using multiple whaleboats, but for some reason moving the whale was impossible this time. Possibly the ice was broken up and there was no clear path back to the ship.[126] By abandoning this whale, Tyson and his crew were losing a substantial portion of their wages for the entire voyage. To get some return for their labor, the men began hacking out the whalebone from the whale's mouth. Known today as baleen, it is the thick black fingernail-like material that hangs down in slats from the upper jaw of nontoothed whales. Living whales use their baleen as a food strainer, scooping up tons of seawater and tiny plankton organisms and then allowing the water to drain out between the slats. After filling their empty boat with a small fraction of this whale's baleen, Tyson and his shipmates abandoned the immense carcass and hauled their boat and its cargo up onto the ice. Already exhausted, they dragged their load all the way back to the ship, where they or others would have painstakingly split, cleaned, washed, polished, and dried the whalebone before stowing it away in the hold. That, wrote Tyson, "is what I call a fair day's work."[127]

The baleen and blubber were the bowhead whale's only saleable parts. They played many of the same roles as today's petroleum products and were similarly ubiquitous. It is hard to imagine a nineteenth-century Qallunaat American home without some trace of a whale inside it. Baleen was strong, flexible, and when heated became a kind of natural plastic that could be reshaped at will. Most famously, it was formed into corset stays, whips, canes, and umbrella frames. A dealer's advertisement listed fifty-three lesser-known whalebone products, including "tongue scrapers, divining rods, plait raises, shoehorns, billiard cushion springs" and "probangs," which were flexible rods "used especially for removing obstructions from the esophagus."[128] In 1850, the whale fishery also supplied most of the nation's industrial grease and light. One American whaler described his job as providing "light to the eyes" and "lubrication to the joints of [the nation]."[129] The oil rendered from bowhead blubber was a relatively low-grade lubricant and illuminant, especially compared to sperm whale oil and spermaceti, which oiled watches and sewing machines, and were used in high-end cosmetics and candles. Still, bowhead whale oil lit the homes of Americans who could not afford more expensive alternatives. It was an important lubricant for heavy machinery and was therefore key in producing many factory products. It was also used to make soap, paints, and varnishes. Whale oil's market share dropped steadily after 1850, with the discovery of coal oil and better processing techniques for lard and tallow, but demand remained high because America consumed so much oil.[130]

A hallmark of the Qallunaat industrial age was a growing distance from the messiness that went into creating all the varied comforts of home—and the exploitation that was enabled by this distance between producers and consumers. Americans who bought whale products were not required to give a second thought to where they came from. But there was nothing easy or detached about hunting and processing whales. Usually, whalers spent several days after a kill wading through the whale's gore. Often, everything but the baleen and blubber was left to rot, although the meat and skin were edible. Inuit salvaged as much as they could, and sometimes returned to whale carcasses months later for dog food. Qallunaat occasionally sampled the meat but did not incorporate it into their regular diet. One crew tried fried bowhead meatballs and pronounced them "tolerable good eating." Another whaler cooked a bit of the shoulder, and commented, "whale steak is good at a pinch." Much of the meat spoiled; bowheads were too huge, and presumably there were too many whales being killed.[131]

The scale of whalers' work can be hard to imagine. A single bowhead could be more than half the length of a whaling ship.[132] Thick, purplish-black blood stuck to the men's skin, matted their hair, and soaked into their clothing. It pooled on the ice and poured into the cold waters of Cumberland Sound. Sometimes, to remove the baleen from the jaw, the men simply climbed into one of their whaleboats and floated into the cavernous mouth. Extending back about sixteen feet, the mouth was as long as some forecastles and at least double the height. Several men could easily stand upright in it while they cut away the baleen from the upper jaw.¹³³ To remove the blubber, the men "cut in" through the whale's skin and peeled off the fat in long, quivering, blanket-like sheets as much as two feet thick. Kowjakuluk, an Inuit Elder, recalled in the 1980s that whalers drove nail spikes into pieces of wood and tied these boards to their boots, providing a sort of crampon that enabled them to walk around on the whale without slipping.¹³⁴ The Americans then chopped up the blubber and minced it into small enough pieces to throw into their smoking try-pots, where the fat was melted down into pure oil. Most of it was packed away into casks, but one whaler reported frying doughnuts in the fresh oil.¹³⁵ Everyone and everything near the try-pots became slimy with smoky grease.¹³⁶ When Americans lit their lamps, or purchased factory goods made on a machine lubricated with whale oil, few would have imagined this scene.

Professional whalers had to inure themselves to the sounds and smells of the hunt. They learned, to some extent, to view the whale as an enemy to overcome to feed themselves and their families, as a form of money that surfaced from the depths. One whaler referred to a bowhead that got away as an "80-barrel whale," estimating how much oil it would have produced. George Tyson referred to the whale his crew killed at the floe edge as "a large and valuable balleener."¹³⁷ Some did question why they chose to spend months away from home doing hard and dangerous labor. "Here I am abstaining from the greatest joys of life just for the sake of amassing a little wealth. What a foolish notion," wrote Ambrose Bates on his thirty-sixth birthday. He went on to say, however, that he hoped to provide his son with a good education—and he did not quit whaling after this voyage.¹³⁸

This does not mean that whalers killed dispassionately or that the hunt had no meaning to them beyond wages. On the contrary, for career whalemen, chasing whales was deeply tied up with their very identity and breath. The hunt gave meaning to their lives and, every time, threatened to take those very lives away. The danger of the hunt likely contributed to the lack of pity expressed for bowhead whales in the records. Whalers, like many of their American contemporaries, not infrequently expressed regret or sadness over the killing of smaller animals, such as seals or hares. George Tyson was devastated when another whaler stomped on a baby seal he had taken in as a pet.¹³⁹ But if whalers encountered a bowhead without the means of killing and processing it, some tried to hurt it anyway.

Ambrose Bates, the first mate of the *Milwood*, wrote about one such incident shortly after it occurred in 1868. It haunted him so much that he had "neither slept nor thought of anything since." Making a tour of Cumberland Sound with a young Inuk as his guide, Bates headed out across the ice with neither boat nor crew. At the floe edge he saw whales. Not just a few but "armies of whales." Bates stared at them, growing increasingly frustrated as they nosed up to him—tauntingly, imagined Bates—as if to say "look and weep." Growing angry, Bates took out his pistol and ineffectually shot one of them four times, while kicking another with his foot. Among the bowheads were smaller beluga whales, whistling loudly. Bates hollered back in frustration. He wrote, "It seemed that it would have been some consolation to me had I had a lance that I might [have] killed a few whales even though there was no hope of saving them."[140] Months later, on his voyage home, he wrote, "I do believe that if I could just lower a boat and attack the ugliest whale that ever swam salt water I should not only conquer the said monster but I should then come to the brig altogether changed."[141] For Bates, whaling had become a kind of personal war, and the whales were commodities that served his considerable personal ambition, and perhaps offered a promise of transformation in his unhappy life.

Since the Qallunaat commercial whalers worked for profit rather than sustenance, their killing was limited only by their abilities and the number of whales. In the 1850s and 1860s, the American demand for whale products was virtually without bounds. The number of whales was not. By the late 1860s, after less than three decades of commercial whaling, Cumberland Sound had been "fished out." The bowhead whale population had declined enough that most captains were looking for better whaling grounds in Hudson Bay and the Western Arctic. By moving through various whaling grounds, American whalers continued to turn profits from bowhead whaling voyages until demand for their products fell away. A few ships continued to visit Cumberland Sound, particularly around the turn of the twentieth century, when whalebone prices were high.[142] The American whaling industry had treated Cumberland Sound as a rather unfortunately situated repository of whalebone and blubber, not as a hospitable place that could sustain people year-round and for generations to come. This commodity-focused outlook had real consequences for the land and for Inuit, who remained in Cumberland Sound and learned to live in a changed world. It also resulted in much American suffering and death.

★

There is no denying that Americans assumed considerable risk by signing onto an Arctic whaling voyage. When they set out from their New England ports, they could expect that at least one person on their ship would never return home. Yet the biggest risks American whalers faced in Cumberland Sound—contagious disease, malnutrition, and horrific accidents on the job—were dangers familiar to laborers in the United States. Nineteenth-century workers protested workplace hazards and fought for a living wage, but these risks were also normalized in American society to some extent. Many working-class Americans in this period consumed a cheap monotonous diet and did unsafe jobs for little pay.[143] When American whalers recorded their fears in Cumberland Sound, it was not the dangers and difficulties of the work that inspired terror, but rather the unfamiliar location: its perceived bleakness, its distance from their loved ones, its lack of forests and farmland and wooden houses. Few whalers who survived chose to sail north again; for most of them, Cumberland Sound never became home. This is partly because they—and their employers who paid and fed them inadequately—so often imagined the Arctic as a trial that could lead to future comfort in the United States, not as a place where they could learn to be comfortable.

But Arctic whaling ships were still a sort of home. Recall the final words of the *Andrews*'s logbook after the shipwreck: "So ends this voyage and a good home." Little has been written explicitly about ships as home places; scholars more often contrast ships with the homes that male sailors left behind. But the literature has too uncritically mirrored the recent association of home with private family space. Home has always been more than this. Some Arctic whalers refered to their ships as homes, and they became familiar spaces.[144]

Geographers have written about how life at sea is "ship-shaped," meaning that sailors' thoughts, actions, and identities are affected by living on a bounded floating vessel.[145] But if whalers' lives remained "ship-shaped" in Cumberland Sound, they suffered. Ships provided a breakwater with the unfamiliar environment, but whalers needed to make connections outside of their ships to be comfortable. Their quality of life generally improved if they could make home extend beyond the walls of their ship and beyond the confines of their memories. Cumberland Sound whaling ships, like any homes, were never simply dwellings but networks of relationships with the land, animals, weather, and people that surrounded them. As always, these relationships were not without consequences. Whaling ships offered new opportunities and connections for Inuit, but the ships also brought hardships with them.

These networks that surrounded whaling ships appear most clearly with the men who made careers as Arctic whalers. While the whaling industry had little loyalty to specific places, experienced captains did. They knew that a voyage could be made or lost on their knowledge of a specific whaling ground and their relationships with the people who lived there. They traded for local food and clothing. At least from their perspective, they developed close relationships with Inuit men and women, which they relied on and rekindled every time they sailed north. They were willing to forgo some profits to overwinter more comfortably rather than just dream of comforts back home. Their more complex feelings about Cumberland Sound are worth exploring briefly here.

Captain George Tyson worked in an iron foundry in New York City before signing onto the *McLellan* as a greenhand. His published memoirs claim that he was always drawn to Arctic exploration, but his manuscript says more simply, "At the age of 21 becoming disgusted by shop labor and my health failing, I concluded to try the sea." After his first overwintering in Cumberland Sound as part of the *McLellan* crew, he was "anxious to get home," and vowed never to go to sea again. But he again tired of "the stifling atmosphere of [the] iron-factory." He signed on to another voyage and eventually worked his way up to captain. In other words, Tyson was not immediately impassioned with Cumberland Sound, but he apparently liked Arctic whaling better than his other options.[146] After two decades of whaling he was hired onto the *Polaris* expedition and then survived the six-month drift on the ice floe. When he put into Cumberland Sound the following summer, he noted: "This seems like home—it is my old whaling-ground, and here we are, snug and comfortable, in Niountelik [Naujaaqtalik] Harbor, so familiar to me."[147] Even though Tyson had experienced stereotypical "Arctic horrors" on the ice floe, he retained a strong attachment to this specific Arctic place, and he knew it well. Yet Cumberland Sound did not always "seem like home" to him: a few years later he complained of being "isolated from all the world" there.[148]

It was common for career whalers' feelings to remain fractured and conflicted. In some ways and at some times, Cumberland Sound was a home for these Qallunaat, but they always felt the pull of their home communities in the United States. Nearly all whalers returned there when they retired. Still, the Arctic remained part of their sense of home. John Orrin Spicer, who spent time in Cumberland Sound and Hudson Bay most years from 1863 to 1892, had an archway in his Connecticut garden made from "the jawbone of one of the largest whales ever caught." He also reportedly wore his *kamiik*, or sealskin boots, around town, as well as a small vial of seal blood that he said lessened the pain of his multiple heart attacks. He had a walking cane made of antler and baleen. George Comer, a whaling captain in Hudson Bay, had a musk-ox head mounted on his front porch, port and starboard lights on his house, and wore a captain's hat at home. In Groton, Connecticut, Sidney Budington and his wife hung portraits of Inuit on the walls of their sitting room, most likely of Inuit they knew who had visited New England on whaling ships and been photographed there.[149] Sidney Budington and many other Qallunaat were photographed wearing Inuit clothing they had acquired in Cumberland Sound. It would be easy to dismiss these actions as Qallunaat using Inuit culture and northern experience to gain status or attention. Yet I think it goes deeper: for everyone, especially as we age, home is not just about the places and people that surround us; it is also about memories and passing on knowledge. Home is in the past and future as much as the present.[150]

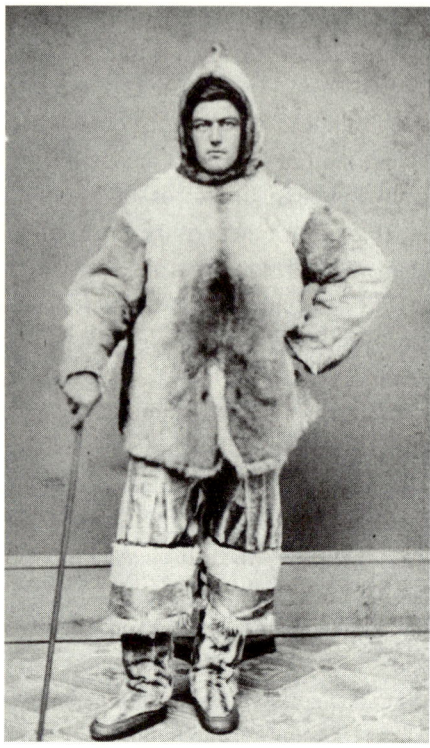

FIGURE 1.5. Whaling captain Sidney Budington in Inuit clothing, ca. 1860. Coll. 257, Manuscripts Collection, G. W. Blunt White Library, Mystic Seaport Museum, Inc.

Although it was not generally talked about, many Qallunaat must also have missed Inuit partners and children—their other families they had left behind. Whalers' experiences in Cumberland Sound presumably changed them in additional ways that they did not record and that we do not know. It also changed the lives of Inuit in ways that we cannot fully know, as this period of intensive Qallunaat whaling is beyond living memory.

Only one known Qallunaaq whaler, William Duval, settled permanently in Cumberland Sound. After bowhead whaling ended in the early twentieth century, he lived into old age with his Inuit wife and children. He died and was buried, according to his wishes, at his home of Usualuk. His Inuit name Sivutiksaq, meaning the harpooner, is still alive in Cumberland Sound today. He continues to be at home there.[151]

2

Inuit in the United States

On August 9, 1862, a young Inuit couple set sail for the United States. Hannah and Ipiirvik were living fifty miles south of Cumberland Sound when the explorer Charles Francis Hall showed up with a whaleboat crew. The couple—also known as Joe and Hannah, or Ebierbing and Tookoolito—already knew Hall well.[1] They had been working for him for nearly two years: guiding him, interpreting for him, sewing his clothes, sharing their knowledge, and often living under the same roof. Now Hall was going home, and as he had mentioned several times before, he wanted the couple to sail to America with him. Hannah and Ipiirvik agreed, reported Hall, "after some conversation."[2] We don't know what was said, or to what extent the Inuit may have felt coerced or intimidated into traveling. Hall recorded one objection: they were worried their infant son would die on the voyage.

Ipiirvik and Hannah quickly packed their belongings. Along with their baby, and a sled dog, they climbed into Hall's open boat and headed for the whaling ship *George Henry*. The next morning was too calm to set sail. The Qallunaat whalers lowered their whaleboats and towed their ship down the bay, toward home—and the news that their country had fallen into civil war. Inuit families saw them off by circling the *George Henry* in kayaks and *umiat*, larger boats that could carry entire families. Finally the Inuit called out *tavvauvusi*, farewell. Ipiirvik would not see his homeland again for over a decade, and Hannah would never return.

What did leaving home mean to Hannah and Ipiirvik? American observers often imagined that the couple's lives improved on leaving the Arctic, but in many ways this was not true. Surviving records suggest that they struggled to adjust to a new landscape, climate, and ways of thinking. For Inuit travelers,

"*Ter-bou-e-tie, In-nu-it*"—(Farewell, Innuits).

FIGURE 2.1. Farewell scene when the *George Henry* sailed south in 1862. Hall, *Arctic Researches*, 563. Image courtesy of the University of Calgary Special Collections.

life in the United States could be as strange, as lonely, as confusing, and as dangerous as American encounters with the Arctic world.

This chapter is a counterpoint to the previous one. I examine Inuit attempts to adjust to life in America, with the goal of showing that it was no more or less a "natural home" than Cumberland Sound. In both places, culture and environment intersected to make it hard for outsiders to feel at home—but in very different ways. Hannah and Ipiirvik had some things in common with career whalers: they spent much of their lives abroad, and developed ties to people and places in both the Arctic and the United States. But their experiences were different. Whalers were backed by a massive industry intent on extracting resources. They arrived with shiploads full of supplies, at least in theory enough to survive the winter. Inuit traveled light and were vastly outnumbered. They came with no food or shelter, and they were dependent on Qallunaat for their transport home. While building relationships with local people was important for whalers, it was essential for Inuit. Five factors that hindered Hannah and Ipiirvik from feeling at home will be discussed in more detail below: constant scrutiny and judgment by strangers, frequent travel, authoritarian attitudes toward them, a new type of economy, and devastating illnesses.

Ipiirvik and Hannah were among over two dozen Inuit who visited the

FIGURE 2.2. Signed portrait of Ipiirvik, wearing skin clothing presumably made by his wife. Nourse, ed., *Narrative*, 1879. Image courtesy of the University of Calgary Special Collections.

United States in the mid-nineteenth century, mostly as passengers on Arctic whaling ships. Like American Arctic whalers, Inuit generally expected to overwinter once and then return home. Their reactions to America varied, as did their reasons for traveling, which likely ranged between a desire to visit the homes of Qallunaat friends, to acquiring valuable trade goods or language skills, to feeling pressure to comply with Qallunaat requests to travel. It is impossible to know exactly how many Inuit went abroad, since they are often confused with each other in the newspaper accounts, or called by widely varying names, or possibly no longer in any surviving records.[3] In any case, the number of Inuit travelers was minuscule by northeastern American standards, but given the smaller Arctic populations and the fact that many visitors came from the same sites, it was enough to shape a culture of overseas travel in places like Cumberland Sound, at least among certain families. Hannah and Ipiirvik had already been to England from 1853 to 1855. They both had relatives

FIGURE 2.3. Signed portrait of Hannah, using her anglicized Inuktitut name Tookoolito. Nourse, ed. *Narrative*, 1879. Image courtesy of the University of Calgary Special Collections.

who had traveled abroad, and the Inuit who saw them off on the *George Henry* included at least one man who had been to the United States and another who would later make the trip.[4] Ipiirvik and Hannah had also become familiar with Qallunaat culture long before going abroad: they had grown up around commercial whalers and had reportedly met the crew of the *McLellan* voyage in 1851–52.[5] Their sense of home in the Arctic probably included relationships with both Inuit and Qallunaat. When they sailed south in 1862, they were not naive about what an overseas trip would entail.[6]

Ipiirvik and Hannah stayed abroad far longer than most international Inuit travelers of the period. They became indispensable to the explorer Charles Francis Hall, both in his fundraising efforts in the United States and on his three Arctic expeditions. Thanks largely to this couple, Hall's enduring trademark as an explorer would be living "among the Esquimaux." Hannah's letters to her American friends, in conjunction with dozens of newspaper articles, diaries, letters, and observers' accounts, allow for a remarkably detailed reconstruction of the couple's years in America. Nevertheless, as with all colonial encounters, much was lost in translation and can never be recovered. Nearly all the words that Ipiirvik and Hannah spoke to each other in Inuktitut have been lost.

To reflect one of the fundamental ways in which Hannah and Ipiirvik reordered their lives in the United States, this chapter is structured around a series of Sundays. In Cumberland Sound, work schedules and routines were established but flexible; they were pegged to variable factors like the arrival of whales and whaling ships, the period when caribou or seal pelts were at their prime for tents and clothing, or the time when the ice became thick enough for travel by dog team. Moons, tides, stars, the sun, seasons, and the movements of animals provided a basic framework, but Inuit also constantly assessed the weather and adjusted their plans accordingly. In contrast, in Qallunaat cities, days and weeks marched on in regular, numbered succession, with engagements written in advance on a calendar.

Hannah was probably already aware of the Qallunaat concept of weeks. She sprinkled the term liberally through her early letters, as in, "How do you get along last eight weeks?" or "Last winter, when I was sick . . . you take care of me many weeks."[7] In Cumberland Sound, the gradual entrenchment of the week as a key unit of time brought many changes. Inuit came to whaling ships and stations to receive weekly rations in exchange for labor; Christians would later abstain from hunting and sewing on Sundays; and many Inuit eventually obtained Monday-to-Friday jobs that conflicted with hunting opportunities.[8] It is far less clear exactly how weeks—and American notions of time more generally—impacted Ipiirvik and Hannah's lives in the United States. This is partly because their own written impressions of time are sparse and partly because American time was quite diverse. Factory workers had regimented workweeks, while farm work was still primarily tied to seasons and daylight. The ways that Americans experienced time were changing in this period, with the construction of railroads upending travel times, a growing urban population working set hours, and the relative affordability and availability of personal clocks. When they worked as entertainers, Hannah and Ipiirvik's days were scheduled by clock time, but irregularly. Sometimes they worked in the evenings or on Sundays, while other times they went to church, and they enrolled their daughter in public school and Sunday school.[9] They certainly noticed the new seasons—particularly the summer heat—but they left little evidence for how the climate shaped their thoughts or actions. What is clear is that in the United States, Hannah and Ipiirvik's lives were no longer chiefly regulated by ecological time, and that American markers of time were no more universal or natural than Inuit ones.

Under Scrutiny: Barnum's American Museum, Sunday, November 16, 1862.[10]

November 16, 1862, was a busy Sunday afternoon at Barnum's American Museum in New York City. Thousands of customers funneled into the brightly painted museum, which sprawled out in both directions from the corner of Ann Street and Broadway. A large cotton banner of an Arctic scene hung outside, and similar images peppered the city in handbills, posters, and newspaper advertisements. They depicted Barnum's latest headline attraction, a living group of "Esquimaux Indians." The museum's owner, Phineas Taylor Barnum, with typical showmanship and hyperbole, declared this Inuit couple and their son to be "beyond all controversy, THE GREATEST CURIOSITY IN THE WORLD."[11]

The six floors of the museum were crammed with Qallunaat patrons of all social classes, and cluttered with thousands—Barnum claimed hundreds of thousands—of curiosities. Cages and aquariums were stocked with shimmering tropical fish, boa constrictor "monster snakes," performing bears, and bovines declared to be "the sacred bull and cow of the Hindoos." Crowds gathered around a packed Happy Family cage, a popular attraction in which predators and prey were forced to coexist, supposedly without eating each other. From 1861 to 1865, Barnum also exhibited a string of unfortunate Arctic *qilalugait,* or beluga whales, but the latest of these had quickly perished in its glass tank on the second floor. In addition to living animals, the museum's dis-

FIGURE 2.4. Barnum's American Museum at the corner of Ann Street and Broadway, New York City, 1859. Museum of Fine Arts, Boston. Edward Jackson Holmes Collection, 65.3233. Published by London Stereoscopic Company. Stereocard, albumen prints mounted to card stock. Photograph © 2017 Museum of Fine Arts, Boston.

play cases, floors, walls, and halls held such authentic and charlatan miscellany as autographs, paintings, historical artifacts, trick mirrors, suits of armor, and wax figures. There were also numerous skeletons and taxidermied animals, supplied in part through the steady stream of deceased menagerie inhabitants. As usual, Barnum was featuring several live human exhibits, most of whom he marketed as controversial figures who bent societal notions of race and normalcy. That Sunday in 1862, Barnum had engaged the dwarf Commander Nutt as well as the so-called Madagascar Albino family, whom Barnum claimed were African but were actually Danish.[12] Amid this heat and clamor drifted Ipiirvik and Hannah and their baby son, all dressed in fur clothing.

New Yorkers who paid twenty-five cents to enter Barnum's American Museum could expect a close encounter with the Inuit. For at least part of the day, the family and the other human exhibits were neither up on a stage nor cordoned off from the public. Instead, they wandered the "crowded salons" of the museum, where they would have been constantly surrounded, jostled, and ogled by patrons. They may also have sat on podiums on the ground floor.[13] In the museum halls, people could have easily approached them, talked to them, touched them. The family's schedule was grueling: they were on display over seven hours a day, seven days a week.[14] Here, in a single day, they would have encountered more people than they saw in a year in Cumberland Sound. Every day except Sunday, they also appeared onstage after the theatrical performances in the lecture hall, an ornate cavernous space that could seat three thousand people.[15] What did it mean to perform difference, while at the same time trying to understand and fit into a new place?

Hannah and Ipiirvik's time at Barnum's museum was an extreme form of the everyday scrutiny they encountered in the United States. Americans were constantly judging the couple as different and inferior, as unintelligible and strange and antimodern. Ipiirvik and Hannah were reminded over and over again that they were foreign. They pushed back against misconceptions and prejudices when they could; this was part of the difficult work of making America their home.

There was little opportunity for resistance at Barnum's museum. Most people who encountered the Inuit there had time for nothing but a quick gawk, a swift passing of judgment, a short anecdote to relate to friends. Hannah, Ipiirvik, and their baby became a canvas onto which museumgoers could indulge their curiosity about the larger world and project their fears, beliefs, and desires about American culture. In a city that was constantly being torn down and rebuilt, that was absorbing thousands of new immigrants a month, the Inuit supposedly offered contact with a frozen and unchanging world.[16] While the American obsession with Inuit culture was relatively new, Inuit were slotted into existing stereotypes about other Indigenous peoples in North America, with some specific descriptors added.[17] According to nineteenth-century reports, Inuit visitors were instantly recognizable by their abnormal appearance, clothing, behavior, and habits. They were as short as children; they dressed entirely in furs and skins. They either lacked emotion or were overly emotional. They exchanged their wives. They consumed pounds of butter while sitting on the floor, and it was "unsafe to leave oil cans about the house." Such descriptions—often exaggerated or entirely made up—served to highlight what kinds of behavior, appearances, and gender roles were acceptable in American society.[18]

Inside the museum walls, Hannah and Ipiirvik seem to have been under considerable pressure to portray these stereotypes and to look like their caricatures in the advertisements. Despite the stifling, stuffy heat of the museum, they appeared in fur clothing. Newspaper articles for other public appearances called it their "Arctic Costume," a "costume peculiar to their country," or a "most outrageous *tout ensemble.*"[19] Americans had little interest in paying to see Inuit wearing cotton dresses or wool trousers. When Hannah showed up for one interview wearing a calico gown, the reporter made sure to mention that at the upcoming lecture, she and her family would be introduced "in their native costumes."[20] Charles Francis Hall stressed when booking at least one engagement that the Inuit would be wearing furs and skins, and the Inuit probably expected this, having appeared in public venues on their trip to England.[21] It was largely this clothing that made Ipiirvik and Hannah seem exotic enough to be worth the price of admission. In a museum full of strangers, and in a vast anonymous city that increasingly relied on visible markers like apparel to denote status, fur clothing allowed Barnum's patrons to easily recognize and judge the Inuit.[22] As one reporter opined, were it not for his "North Pole uniform," Ipiirvik could have been mistaken for an Italian immigrant.[23]

Caribou suits, when sewn by a skilled Inuk seamstress like Hannah, could keep the wearer warm and comfortable in some of the coldest weather on the planet. But they were not designed for Barnum's American Museum. Hall, who had arranged their contract, recognized and pitied the couple's discomfort at appearing day after day "in hot furs & in hot rooms."[24] Fur clothing marked the Inuit as anachronistic curiosities, out of time and place, and certainly not at home in the United States. Worn inside the hot museum, it encouraged visitors to see the Inuit as alien, stolid, unchanging, and perhaps even dim-witted. Caribou-skin attire belied the fact that Hannah and Ipiirvik had been encountering large groups of Qallunaat since childhood, and that they had already demonstrated an exceptional ability to live everywhere from snow houses to whaling ship cabins, from sealskin tents to English homes.

FIGURE 2.5. An advertisement for Hannah, Ipiirvik, and Tarralikitaq's appearance at Barnum's American Museum. The stereotypical Arctic image was literally produced from a "stereotype"—a metal plate purchased from a printer and used in promoting various public appearances. Charles Francis Hall Collection, Archives Center, National Museum of American History, Smithsonian Institution.

The Inuit also seem to have spoken exclusively in Inuktitut while at Barnum's museum. According to reports in the *New York Times,* they conversed only in "uncouth gibberish" or "non-understandable gibberish." One article reported Charles Francis Hall acting as their interpreter.[25] This was illogical and unlikely. After two years in the Arctic, Hall still needed Hannah to translate conversations for him. Her spoken English was so strong that the first time Hall heard her, he mistook her for a Qallunaaq lady.[26] Hannah was proud of her language skills and seems to have enjoyed surprising foreigners by addressing them in English. Ipiirvik's English was weaker, but he was still capable of carrying on lengthy if stilted conversations in his second language.[27] Furthermore, while the couple often referred to each other as "Joe" and "Hannah" when speaking or writing to Americans, at the museum they were known by their "unpronounceable" foreign names.[28] By being presented as unilingual, the couple would have presumably been subject to countless jeers and erroneous statements by visitors who assumed they could not understand. Many Qallunaat also imagined that Indigenous languages like Inuktitut lacked the vocabulary for expressing abstract or complex thoughts.[29] Without using English, Hannah and Ipiirvik had no opportunity to convince English speakers that they shared the same human depths of emotions and intellect—or more precisely, equal but different depths, as Inuit and Qallunaat knowledges and emotional expressions varied greatly. It is probable that Barnum, who routinely created identities for his performers and directed them to behave in ways that played into existing pseudoscientific theories about race, instructed the Inuit to speak only in their own language.[30] Visitors could then encounter them as unintelligibly foreign and feel satisfied with a brief encounter that confirmed their stereotypes.

If this was the charade, it seems to have succeeded. Visitors believed they were seeing authentic people, uncorrupted by artifice. They described the Inuit as "curious specimens of humanity" or as "three of the queerest-looking specimens of humanity this side of the North Pole."[31] The museum was notorious for fraud and humbuggery, but the *New York Times* urged its readers to go see the Inuit, because unlike many of Barnum's living curiosities, they were "genuine."[32] Presumably this meant that they looked and behaved as the reporter had imagined they would. Inarguably, the Inuit were genuine in a way that albino Danes masquerading as Africans were not. But their most "authentic" markers of appearance and behavior were exaggerated, fabricated, and performed.

In Cumberland Sound, Hannah and Ipiirvik had been on their way to becoming *inummariit,* true Inuit. To Inuit, at least in today's living memory, authenticity is a process rather than a birthright; it is the path of becoming tolerant, generous, adaptable, persistent in the face of difficulty, and respectful of the traditions and stories of one's ancestors. These skills were learned through living off the land; being *inummarik* is deeply connected to being at home in one's surroundings. Successful *inummariit* help to build the Inuit world through their actions and relationships to it, through their travels across their home territory, and through their participation in the cycles of hunting and redistribution of wealth.[33] A century later, the Inuit artist and writer Alootook Ipellie was sent south for school, and he wrote that even after returning to his family, he was never again able to "pursue [his] traditional culture and heritage as an Inummarik."[34] He could not truly go home again. By leaving Cumberland Sound, and distancing themselves from its networks of people and animals, Ipiirvik and Hannah were, I think, also distancing themselves from becoming *inummariit*. Certainly they never ceased to be Inuit—and they had plenty of opportunity to practice traditional Inuit values of tolerance and patience in the United States. But performing as "genuine Esquimaux" indoors at Barnum's American Museum was in many ways the opposite of becoming genuine Inuit, or *inummariit*.

Hannah and Ipiirvik had plenty of opportunity to observe "genuine" Americans. They traipsed through and presumably puzzled over Barnum's museum. They left no record of what they thought of the captive and mistreated animals or of their interactions with the patrons and the other humans on display. We can infer that they were unhappy. Even Charles Francis Hall, who was constantly on the verge of penury and appropriated most of the Inuit family's income for himself and his expeditions, stated that the Inuit must never again be "subjected to such trials." He refused Barnum's offers to engage the family again.[35]

Still, Hall kept asking the Inuit to perform in public. Two days after the end of their engagement with Barnum, they boarded a train for Boston, to complete a contract Hall claimed he regretted having signed. The family spent the next two weeks on display at the New Boston Aquarial and Zoological Gardens.[36] Billed once again as "The Greatest Curiosities in the World," they appeared alongside "two gigantic Japanese salamanders," "a superb black African ostrich," a pair of beavers, rebel battlefield relics, an aviary, and yet another "immense Happy Family cage."[37] With no other humans on display, the Gardens implicitly equated Inuit with the animal world. Ipiirvik and Hannah may not have felt demeaned or humiliated by this in the same way many Qallunaat would. However, they would have seen, and probably been appalled by, such behavior as a trainer repeatedly poking a bear with a stick.[38] It could have been a reminder of the foreignness and potential danger of Qallunaat culture. Inuit who unnecessarily tormented an animal were taught to expect direct and visceral repercussions. As the Elder Pauloosie Angmarlik phrased it in the twentieth century, "We were told when we were children that if you mistreated or made fun of animals, they could seek revenge, especially the ones that could be fierce."[39] Following this engagement, Hall signed no more contracts to exhibit the Inuit in museums, aquariums, or other flashy establishments of the period.

Hall still expected the Inuit to appear at his lectures. Following in the footsteps of the revered American explorer Elisha Kent Kane, Hall gave numerous public talks on an established lecture circuit. He hoped to attract wealthy patrons and cover his own living expenses. Unlike many Victorian explorers, Hall had no scientific training and professed little interest in the natural sciences. He was a middle-aged newspaper editor in Ohio who had become obsessed with the search for Sir John Franklin. Hall never connected very well with the scientific community, but he did appeal to popular audiences, and the Inuit helped.[40] They frequently appeared alongside him, in a far more sedate environment than the museums but still under intense scrutiny.

Newspapers reported that "all other attractions of the evening were thrown into the shade" whenever the Inuit family took center stage at Hall's lectures. They "created quite a sensation" and "every eye in the audience . . . turned upon them with interest and pleasure." On at least one occasion, Ipiirvik and Hannah "held a *levee* of some duration, until they were almost deprived of breath." Sometimes, audience members stayed around long after the end of the lecture, simply to observe "the interesting family."[41] Hannah and Ipiirvik's own conceptions of family and kinship would have extended far beyond this tiny group of three. To American audiences, however, their father-mother-child unit likely served to support an emerging middle-class assertion that nuclear families were universal and natural.[42]

If the Inuit ever enjoyed aspects of performing in public, they must have tired of it. In later years, when they were living on their own and in need of money, I have found no records of them choosing to appear in front of crowds. In their everyday interactions with Qallunaat, they strived to be recognized as intelligent, adaptable, generous, and relatable human beings, rather than as "THE GREATEST CURIOSITIES IN THE WORLD."

Outside of museums and lecture halls, there are many accounts of Hannah, Ipiirvik, and other Inuit visitors not behaving as Qallunaat expected. Most Americans were looking for strangeness, but Inuit often sought to appear as unremarkable as possible. They were "quick to learn" and "endeavored to do as other people did." Observers were fascinated by the Inuit ability to imitate Qallunaat manners and customs, and to absorb unusual sights without displaying any outward signs of surprise. When the visitor Kudlago saw a locomotive engine for the first time, he "expressed no words & exhibited no signs but what were consistent with the fact of his having seen the same a thousand times before."[43] This lack of reaction would have been astounding to many nineteenth-century Americans, who were overwhelmed by the speed, power, and noise of trains. Similarly, when Inuluapik, Hannah's older brother, visited Scotland, he was invited to a fancy dinner party. His hosts hoped that his strange eating habits would provide some entertainment, but they were disappointed. Inuluapik carefully observed the other guests and imitated their table manners so precisely that he appeared to be at home.[44] Americans who heard Hannah speaking English, or who saw the Inuit drinking tea in western dress, likely also questioned notions that Inuit were fundamentally different from themselves. Successful imitation did not, of course, indicate that Hannah, Ipiirvik, Kudlago, and Inuluapik necessarily felt at home in these situations. Their well-being and comfort depended on fitting in and not causing offence.

Hannah and Ipiirvik adopted useful aspects of Qallunaat culture while trying to disprove negative preconceptions about what it meant to be Inuit. They would already have honed skills of close observation and adaptation, by studying, anticipating, and quickly adjusting to changing weather and movements of animals. Change was a constant in their homeland. To a much greater degree than the average Qallunaaq, Hannah and Ipiirvik likely considered it self-evident that they should alter their behavior in a new environment. In other words, by adapting to some Qallunaat customs, I think they were expressing their skills and knowledge as Inuit.

Soon after arrival, Hannah and Ipiirvik adopted American clothing and wore it nearly all the time when they were not performing. Cloth clothing was more practical for everyday living in the United States and much easier to make or acquire. Through Charles Francis Hall's contacts, Hannah soon received cloth dresses. Ipiirvik received similar presents and occasionally shopped for garments. Hannah eventually acquired a sewing machine.[45] One reporter who visited the Inuit disappointedly noted them "dressing and eating as we dress and eat."[46] American clothing allowed the Inuit to blend in, to be seen as potential friends and equals. Most Qallunaat in this period imagined Inuit as primeval, inferior versions of themselves. For Hannah, Ipiirvik, and generations of other immigrants, dressing to conform to local norms served as a superficial but persuasive indication that they were not so different after all.

Hannah may also have tried to push back against wearing caribou clothing at Barnum's American Museum. Charles Francis Hall had brought furs to the United States, and he asked Hannah to sew them into traditional clothing for public appearances. Two weeks before the Inuit appeared at Barnum's museum, Hall urged Hannah to "work as fast as she can" to "make up fine . . .

dresses of the rein-deer skins which I brought." He stressed that the caribou clothing be "*made Esquimaux style*"—a necessary stipulation, since up north, Hannah had sometimes sewn skin clothing modeled after items she had worn abroad in England.[47] In another letter the following week, Hall worried that Hannah was not working quickly enough, and that she would not finish the clothing before coming to New York.[48]

Turning caribou skins into full winter suits is a highly skilled, arduous, and time-consuming task. But still, I think it's likely that Hannah purposefully delayed completing the garments. She would not have been the first Inuk to object to wearing fur clothing abroad. Her brother, Inuluapik, overheated while wearing his skin clothing for a kayaking demonstration on a warm day in Scotland. He had only agreed to wear it on the condition that he could put it away until he returned to the Arctic.[49] Hannah was by all surviving accounts a polite and gracious woman, who seems to have respected both the nineteenth-century American notion of female deference and the traditional Inuit avoidance of direct conflict.[50] She would have been unlikely to refuse Hall directly, and she seems to have finished the clothing. Reporters remarked on the Inuit wearing "the regular costume of the Esquimaux" at public appearances that fall.[51]

Over time, Hannah and Ipiirvik managed to become more than just a stereotype to some Qallunaat. They did not dispel all prejudices about Inuit and the Arctic—that would have been impossible—but they found appreciation as individuals, rather than simply "Esquimaux types." One man who observed them in England repeated many misconceptions about Inuit in his writings, and noted that "the Esquimaux are described by most travelers to be gloomy, morose, dull, stupid persons." But after meeting Ipiirvik, he considered him "an intelligent, quiet man," "grave and serious in his deportments," with an "ease and dignity" to his movements. He described Hannah as "lively" and "quick" at picking up new skills such as English, and with "a joyous laugh." "She is," the writer continued, "a great favourite with all who see her sufficiently to know her." Hall would later describe Hannah as gentle, graceful, "remarkably intelligent," and possessed of a "calm intellectual power."[52] And at least one group of American "fine gentlemen" coveted a pair of gloves that Hannah had made and "wondered that an Esquimaux could do such nice work."[53] This last quote also reveals the level of ignorance the Inuit had to overcome: Hannah and most of her contemporaries were expert seamstresses; their families' lives depended on having comfortable, well-fitting warm clothing.

The couple would never stop being reminded of their foreignness. Like the ambiguous compliment about Hannah's gloves, anything the Inuit did was apt to spark "wonder." Even when they dressed and spoke like Americans, Hannah and Ipiirvik attracted attention everywhere they went. Hall's potential patrons were frequently thrilled to meet Ipiirvik, who sometimes accompanied Hall on his errands. At least once, Hall stopped on the street to give people "the opportunity [of seeing their] first Esquimaux."[54] The scrutiny became even more intense after they returned to Qallunaat society following their rescue from the ice floe in 1873. On landing in Newfoundland, the Inuit were bombarded with visitors, and the children became ill from eating too many gifts of candy. When they spent the following summer in Wiscasset, Maine, they received several hundred visitors a day.[55]

Much of the attention Inuit received—or at least the attention that Qallunaat observers chose to record—was well intentioned. Americans encountering Hannah or Ipiirvik for the first time expressed delight and excitement. They invited the Inuit to come visit them anytime. They asked Ipiirvik to spend the night in their homes. They wrote to Hall, thanking him for introducing them, describing the meeting as a "*red-letter* day" or "a *treat* which rarely occurs."[56] Hannah and Ipiirvik may have been warmed and comforted by this hospitality from strangers, by the invitations into homes, the sharing of food, and the gifts of fine second-hand clothing. Inuit did the same for American whalers who arrived on their shores. But at times, the constant attention was likely intrusive, tiresome, and irksome: a barrier to feeling at home. Hans Hendrik, the other Inuit father on the *Polaris* ice floe, described the summer he spent with Hannah and Ipiirvik in Maine in 1873: "During our stay here in the country of the farmers . . . we also frequently had visitors, who came in large carriages. Sometimes, at noon, when we were going to have our dinner, and people crowded in, we felt embarrassed. However, they were all very kind."[57] Like many of their American counterparts, Hannah and Ipiirvik would not have been used to much privacy, but neither were they used to meeting new people almost every day, to being the constant center of attention, to having every word or gesture judged. One American who spent considerable time with Ipiirvik commented that the man was "naturally of a retiring, modest disposition" and had "a quiet dignity and gravity about him which effectually repels anything like idle curiosity, and resents being regarded in the light of a mere show, which people are only too apt to behold in a poor Esquimau."[58]

Hannah and Ipiirvik must have heard plenty of insulting and ignorant comments about their behavior, abilities, and homeland. Adults who were introduced to the family socially might have been too polite to make negative judgments face-to-face, but what would Connecticut children have repeated to Ipiirvik and Hannah's child in the schoolyard? Some Inuit, even to this day, criticize Qallunaat for being sulky or losing their tempers and for allowing negative emotions toward other people to surface in ineffectual ways. There is no recorded instance of Hannah or Ipiirvik displaying anything but cheerfulness and generosity toward Americans who were eager to meet them. While this was probably not always the most effective tactic for dealing with aggressive Americans, it may have allowed them to remain at home with themselves. I do not know how the couple felt about their Inuit identity throughout their years in the United States, but in their tolerance and restraint, they upheld Inuit values, at least as I understand them today.

Constant Travel: Ferry Terminal, Bank Street, Groton, CT. Sunday, March 1, 1863.

At midnight on March 1, 1863, three dismal figures disembarked at the ferry terminal in Groton, Connecticut. A cold winter rain began to fall; it would continue through all of Sunday. Ipiirvik, Hannah, and Charles Francis Hall had been traveling since Saturday afternoon, and the Inuit couple were both sick. They huddled miserably at the dark terminal on Bank Street, waiting for the whaling captain Sidney Budington to meet them. Budington had given Hall and the Inuit passage on the whaling ship *George Henry* the previous summer, and he remained in close contact with them. When Budington arrived at the terminal, he took the visitors one and a half miles northeast to his house in Pleasant Valley, where they found Sidney's wife Sarah in a state of shock, completely "overwhelmed" with grief. Hannah was even worse; she alternated between periods of anguish, delirium, and unconsciousness. The party carried with them the body of Tarralikitaq, the eighteen-month-old son pictured in the Barnum advertisements. He had died in New York City the previous morning.[59]

The loss of Tarralikitaq was immense to both parents. It was something they had feared from the moment they had boarded the *George Henry*. They had told Hall they were worried that their baby would die on the ship, and indeed, the confined quarters and exposure to germs on board posed a danger. Yet the Inuit were pressured to travel almost constantly after they disembarked in New London harbor, putting themselves and their child at increased risk.

Tarralikitaq had first fallen ill over five weeks previously, just after the family and Hall had returned to New York City from a fund-raising lecture circuit. This tour had taken them to five cities and nine venues in a little over two weeks. Between these lectures, their museum engagements, and their interim stays in New York and Connecticut, the Inuit were more mobile that first winter in the United States than they would have been in Cumberland Sound. The constant transience took both a mental and physical toll on the family. It arguably made it harder for them to feel at home in America and to develop lasting relationships with people and places.

Hannah and Ipiirvik had been traveling since childhood, but most of their movements around Cumberland Sound would have reinforced rather than disrupted their sense of home. Inuit relocated according to seasonal patterns, and most often through sites well known to their ancestors. They returned to the same sites, or at least to the same areas, during the same season each year. Americans tended to hold grossly simplified ideas about what it meant for Inuit to be "nomadic." They imagined that Inuit had little attachment to specific places and thrived on constant movement. R. W. Seager, a traveling entertainer in Ohio, implored Hall that winter to "lend" him the Inuit family, despite knowing all three of them were ill. He argued: "Would not the kind of life I am leading, exciting, active, moving about continually . . . keep them in such good spirits that they could not get sick?" Seager added that he was "something of a Doctor." Tarralikitaq had recently died when Hall received this letter, and it seems he never responded to it.[60]

Hall had previously considered Seager's request, however, and he shared the same misconception about Inuit nomadism in his lectures, when he announced: "They seldom remain permanently in distinct settlements, since there is no agriculture among them, but wander restlessly from place to place, seeking where they may find the best hunting grounds, and thus supply their only wants—food and clothing: therefore the explorer is liable in any locality, and at any time to meet parties of them on their travels."[61] This was wishful thinking on the explorer's part, or perhaps a deliberate attempt to persuade audiences to support his next expedition. Elder Ollie Ittinuaq from Rankin Inlet recently offered a very different view of the time before Inuit were resettled into permanent communities: "People spent the spring in some places and the fall in others. They went back . . . year after year when they went caribou hunting or fishing. This was part of survival in the days when people depended on animals. . . . Whenever we arrived at a site, we always felt comfortable. We were home and we knew where there were animals."[62] Hannah and Ipiirvik's home territory was vast by American standards, but they did not live everywhere; they did not wander everywhere; there were many places where explorers would never encounter them. Family groups did relocate, sometimes over hundreds of miles, for a variety of reasons. These could include seeking trade goods from explorers or whalers, visiting other Inuit, tracking animal migration patterns, or moving closer to a dead bowhead whale—in precontact times, it was often easier to move the community than to tow the whale. Families also moved at least temporarily during epidemics, to allow the land to "cool off." Yet even today, many Inuit who have been resettled in permanent communities continue to hunt in the specific region where they grew up. It has history and significance to them, and it is there that they know the land and the animals.[63]

Hannah and Ipiirvik arrived in the United States at a moment when new methods of transportation and communication were disrupting, enabling, and forever altering lives. In the winter of 1862–63, the Inuit crisscrossed the northeastern United States with Hall by train, at speeds that would have been unthinkable to previous generations. Hall would later call it "this new, living, *lightning-moving-world.*"[64] Many Americans were troubled by the pressure they felt to travel and relocate. When they imagined Inuit were content to be nomadic, they were also saying Inuit were different from themselves.[65]

Nowhere was the transportation revolution more evident than at one of Hall's lecture sites in Elmira, a booming community in upstate New York at the crossroads of two canals and two railroads. Hall does not seem to have known anyone in town, but Ira F. Hart, a local physician and the chair of the local YMCA lecture committee, had read about the Inuit in the *New York Herald.*[66] He had written to Hall, urging him to give a lecture in Elmira, stressing its easy accessibility as a transportation hub, and offering accommodation for him and the Inuit at Brainard House, a new hotel that had been built to coincide with the arrival of the Erie railroad.[67] Hall accepted, eagerly adding two Elmira lectures to his roster.

Sickness was the clearest tangible consequence of all this hurried movement and exposure to crowds. In Elmira, Ipiirvik was the first to succumb to what Hall called a cold but which was likely a more serious pulmonary illness. After completing the Elmira lectures, Hall ushered the Inuit onto a train to New York and through the city streets. He rented an inexpensive suite of rooms on the fifth floor of a building at 33 Bowery, in an impoverished overcrowded district. The suite, if typical for the time, was likely a tiny "parlor" backed by a windowless bedroom.[68] The Inuit and most Americans would have been unaccustomed to living so high above the ground, in densely packed tenements. When Sidney Budington visited the apartment, he wrote excitedly to his wife that he was "5 stores [stories] above the streat."[69]

Budington did not tell his wife that he was staying less than a block south of the raucous Bowery Theatre and along the edge of the Five Points district, then the most notorious poor immigrant neighborhood in New York City. Five Points was safer than it had been in the antebellum period, but President Lincoln was still shocked by it in 1860, and it remained a center of gambling, petty theft, drunkenness, filth, and grinding poverty. Just a few blocks from Hall's rented suite, nearly all the buildings would have housed cheap brothels, and women solicited sex along the Bowery at all hours. In 1863, at the height of the Civil War, there would also have been soldiers and crippled veterans in the streets, and an endless stream of funeral processions.[70]

The Inuit had witnessed hunger, illness, disfigurement, and a type of sex trade with whalers in their home communities, but not with the intensity and concentrated desperation of the Five Points district. Human and animal waste clogged the side streets to a degree unimaginable; there were over twenty thousand horses in New York City, hundreds of which died in the streets every year.[71] Had the Inuit stayed and survived, as countless immigrants did, they might have become accustomed to these streets and made friends among their neighbors. But this apartment was one more stop for them, one that would always be marked by sadness and loss.

Hall spent most of the first night in the room awake, watching Ipiirvik, whose condition continued to worsen.[72] Two days later, Hannah and the baby Tarralikitaq were also "prostrate with severe sickness." Hall was, he wrote, "in the midst of the deepest anxiety and trial." The three Inuit were so ill they could not get out of bed. Hall knew little about caring for the sick, and a fetid walk-up tenement was not the ideal place to learn. On Tuesday, January 27, he wrote in desperation to Sidney Budington, begging him to "come at once" and take the Inuit back to Connecticut where he and his wife Sarah could care for them. Budington came, but the Inuit seem to have been too sick to travel.[73]

The family recovered temporarily, but two weeks later, Hall had to break an engagement because Tarralikitaq was too sick to go out. The toddler was then eighteen months old, and according to Hall, he was usually an "animated, sweet-tempered, [and] bright-looking" child. He charmed onlookers by trying to imitate everything he saw and heard. The boy had learned to run two months previously, but by February 23 he was too ill to walk. Hannah and Ipiirvik's language had a variety of terms for sickness, including *qanimajuq*, a verb used for those like their son who were bedridden and who might die. On February 27, Hall again thought it would be too risky to move Tarralikitaq to Connecticut. The boy died at 8:30 the next morning.

In the days following Tarralikitaq's death, Hall described Ipiirvik as "disconsolate" and "afflicted." Hannah, already seriously ill, became suicidal with grief. Everyone at the Budington household thought that Hannah would not survive long. She was *kipiniaqtuq,* or grieving to the point that her physical health suffered. She was unconscious much of the time, and when she awoke, she often asked for her son, calling out, "Where's my 'Johnny'?" She expressed a desire to die. Her condition worsened after Groton's Presbyterian minister buried the child.[74]

Hannah and Ipiirvik were grieving far from home and other Inuit. Since

Tarralikitaq perished in a foreign country, away from their land and where no other Inuit families lived or knew of their loss, his name, or *atiq*, could not be passed on to a new child anytime soon. The name could not go home. The return of names often softens grief, but the couple would never know their son's name again in a new person.⁷⁵

Somehow, Hannah regained the will to live. Her health slowly began to improve.⁷⁶ She knit gloves for the doctor in New York who had tried to save her son.⁷⁷ She refused to leave Groton, however, and gave the gloves to Hall to deliver. Nearly two months later, Hall visited the Budingtons and watched Ipiirvik paint the fence, while Hannah walked gingerly around the yard. Her legs were weak and she had serious pain in the small of her back. She was recovering physically, but mentally she was still suffering. Hall was trying to persuade the family to go to the Arctic with him again. Ipiirvik may have been willing, but Hall reported that Hannah did not want to leave her son's grave.⁷⁸

The historical record says little about Ipiirvik's opinion of the family's many relocations in the United States, but Hannah's letters to Sarah Budington provide evidence that she wanted to stay in Groton. Between 1863 and 1873, Hannah wrote Sarah whenever they were apart: from New York City; from a farm in Nyack, New York; from Hudson Bay while on expedition with Hall; and from her summer in Wiscasset, Maine. Even though Inuit had no written form of their language in the mid-nineteenth century, Hannah presumably realized that literacy was a skill highly valued in the United States. She spent considerable time improving her reading and writing abilities, sometimes practicing every day.⁷⁹ These efforts helped her develop a close and lasting friendship with Sarah Budington. In their letters, the two women focused on shared issues, especially children, family, and the mourning they experienced following the death of loved ones.

All of Hannah's letters to Sarah evoke sadness over their separation, and often a sense of resignation at having to travel. When Hall took Hannah and Ipiirvik to Nyack the summer after Tarralikitaq's death, Hannah forced herself to get up and sew every day, but she confided to Sarah that she could not escape her sad thoughts and fears. "Down hearted," she wrote to Sarah, "worry and worry . . . my little Jonny I lost."⁸⁰ Hannah was still grieving deeply a year later, when she told Sarah that she often still wept over her dead child: "I cry an cry. Sometime feel better," she wrote. She said she wanted to "go there where little Johnny was," presumably Groton.⁸¹ Hannah seems to have felt comfortable writing to Sarah, who likely empathized. Sarah would also have grieved any lost child openly and at length. Literary critic Dana Luciano argues that middle-class Americans, who previously had tried to stifle grief, now allowed themselves to mourn emotionally, in part to step out of modern linear time

FIGURE 2.6. Hannah. Photo from the Hannah and Joe Ebierbing Collection and reprinted courtesy of the Indian and Colonial Research Center, Incorporated, Old Mystic, CT.

and connect with the sacred and the eternal.[82] When Sarah reported that she had visited Tarralikitaq's grave, Hannah broke down into tears. She urged Sarah to take care of the toys she had left on the boy's grave.[83]

Hannah seems to have been particularly concerned about her son's gravesite. When in Groton, Hannah visited the cemetery regularly, and after her son's death, in accordance with Inuit custom of the time, she had collected Tarralikitaq's toys and placed them on his grave. She was devastated when someone removed a painted tin pail from it.[84] In some Inuit conceptions, the objects on a grave belong to the dead person; alternatively, the essence of the dead person becomes attached to them. Removing property from Inuit graves could invoke supernatural retribution, not just on the perpetrator, but also on his or her family members.[85]

While Hannah was refusing to leave her son's grave, Charles Francis Hall was eager to depart on his second expedition, this time to a new place hundreds of miles west of Cumberland Sound. He remarked that Hannah "does not want to go to her Northern Home."[86] Home is a complicated concept in both English and Inuktitut, linked to ideas of shelter, safety, emotional

FIGURE 2.7. Portraits of unidentified Inuit women with contented babies in their *amautiit* coats, drawn by Hannah before leaving for the United States. Charles Francis Hall Collection, Archives Center, National Museum of American History, Smithsonian Institution.

attachment, identity, and familiar land or territory. Hannah did not, as many Americans supposed, feel at home anywhere there was ice and a midnight sun. She finally went with Hall to this part of the Arctic unknown to her, which she described to Sarah as "nune [*nuna*]," a concept loosely translated as "land" but that also encompasses the idea of living off the land and the sea. Cumberland Sound resident Louee Mike more recently evoked the power of this word and its connections to home: "When I think of *nuna* I think of everything else, like survival, living, future, everything. It's not just the landscape, it's not just how it looks, it's alive for me."[87] Today Inuit in Canada collectively refer to their homelands as Inuit Nunangat. So while Hannah recognized her new surroundings as a place where Inuit lived, she also said it was a "new land, low wind, very cold," and not nearly as warm as "her country," probably Cumberland Sound. She said it was nothing like New York and Connecticut, and that she felt "home sick."[88]

Hannah expressed a stronger sense of home in Groton in her writings than she did in parts of the Arctic unfamiliar to her. Leaving for the *Polaris* expedition years later, she reportedly cried in the carriage as she left Groton.[89] In one letter, she explicitly referred to the Budington house as home.[90] Hannah's friendship with Sarah Budington helped her feel at home in Groton.

FIGURE 2.8. Scenes of Inuit at home in the Cumberland Sound region, drawn by Hannah ca. 1860–62. Charles Francis Hall Collection, Archives Center, National Museum of American History, Smithsonian Institution.

But the death of her son Tarralikitaq also seems to have tied her to the United States. I do not know if she also revisited other sites where her son had lived, if she later stood outside the Bowery apartment or Barnum's museum and thought of him. But she wanted to stay close to his gravesite. Of course this decimating tragedy, this loss of a child, did not make Hannah "feel at home" in the common meaning of relaxed and loving comfort. But home is not just about positive feelings; it is also about rootedness; it encompasses both love and grief. When Hannah buried her son, she lowered a part of herself into the Connecticut earth.

In her letters to Sarah, Hannah makes few direct mentions of Cumberland Sound. A year after her son died, when she was once again ill and bedridden in New York City, she wrote that she was "far from my home. Ice snow."[91] She did not specify what she thought of when she remembered Cumberland Sound, which place or places, or which relatives and friends.

Ipiirvik left very few records behind, so it is much harder to know how he thought about home. But he often talked about his old houses of snow and skin while in the United States. One winter day he put on his fur clothing and made a snowhouse or *igluvigaq* in the yard, with a piece of pond ice for a window, and the family slept in it that night.[92] Types of dwellings are deeply connected to people's sense of home. In Cumberland Sound the Inuit couple would have visited whaling ships, but their primary residences were *qammat* (sod and skin houses), *tupiit* (sealskin tents), and *igluvigait* (igloos). A century later, Inuit resettled into wooden houses would sometimes feel they could not truly communicate inside the new architecture. Building this *igluvigaq* was a way for Ipiirvik to bring his family home. It could have provided a space for him and his wife to speak Inuktitut, using the specialized words that no longer had meaning in their everyday life in the United States. There is no way to know what they said inside the *igluvigaq*, what emotions they expressed, how they felt to wake up and see the snow ceiling above them.[93]

A divided sense of home is a feeling familiar to many immigrants. Homesickness, dislocation, and malaise were also increasingly common among many native-born Americans of the Civil War era, when so much of the population was at war, displaced, or on the move in search of opportunity.[94] Even Charles Francis Hall once reported feeling so "Home-Sick" in New York that he "could stand it no longer." He added that he felt at home at the Budington residence, and with his patron Henry Grinnell in New York, but nowhere else—conspicuously leaving out the Ohio residence of his wife and children.[95] Years later, he would claim, "The Arctic Region is my home. I love it dearly; its storms, its winds, its glaciers, its icebergs; and when I am there among them, it seems as if I were in an earthly heaven or a heavenly earth."[96]

I want to pause briefly on this quote from Hall, as it has been reprinted many times, and I think it is worth examining more deeply. I read Hall's love for the Arctic as sincere, indeed consuming. Hall died on his last expedition; he lived in the Arctic for years at a time; and in America he was ever-fixated on heading north again. Yet as a description of home, his quote is missing two standard threads: the presence of other people, and the land as a source of sustenance beyond the spiritual. This does not reflect Hall's lived experience. When he was in the Arctic, he traveled mostly with Inuit who fed, clothed, and housed him from the land. For much of a decade Hannah and Ipiirvik were part of his home, wherever it was. But Hall's expressed sense of home in "the Arctic Region" does not account for this. It stands in contrast to Hannah and Ipiirvik's statements about parts of the Arctic outside of Cumberland Sound, where they expressed homesickness or distrust of local Inuit.[97] Hall's vision of Inuit society, at least as expressed in his surviving lecture notes, was of a healthy population whose basic needs were warmly met, but where there was little affection within families, few thoughts beyond bare survival, and a general willingness to relocate wherever Hall wanted to go. This is of course indicative of how Hall failed to understand the depth of emotion, thought, and culture in Inuit communities. It is also antithetical to how Inuit and most others envision a home, where relationships and meaning and attachment to place are everything.[98] In short, even when two people feel at home in the same place, it does not mean they understand or value it in the same way, or agree on what is most important. This becomes critical when one party feels they have the right to make decisions for the other, as Hall did with Hannah and Ipiirvik, and as so many Qallunaat have done with Inuit.

To return to Hannah and Ipiirvik, their home in the United States certainly included relationships with other people, most notably Charles Francis Hall. These three lives remained intertwined for over a decade, and Hall played no small part in the Inuit couple's itinerant lifestyle and its attendant emotional and physical traumas. In one of Hannah's first letters to Sarah, after a year of whirlwind travel, she asked the question: "Where is my home?" In that letter, she answered by saying it was Sarah's house in Groton. But I do not think that, throughout her life, she had a simple and constant answer.[99]

Authoritarian Attitudes: Grinnell Residence, New York City. Sunday, June 21, 1863.

The 1863 summer solstice found Charles Francis Hall in a dark and furious mood. He sat in the depths of Manhattan, in a city simmering with resentment over the Civil War draft. Hall was too old for the draft, and he seems to have taken remarkably little interest in the war. His foul mood was due to a personal feud with Sidney Budington. In Henry Grinnell's posh apartment, Hall penned an outraged letter. As if ending relations with a lover, he wrote that he was sending Budington a shawl that Sarah had loaned to Hannah, and that his agent would go to Groton to collect any items the Inuit had left behind. Hall declared that neither he nor the Inuit would ever visit the Budingtons again. "Confident am I," asserted Hall, "that the time will come when you will deeply regret the strange & unaccountable treatment you exercised toward the writer of this & the Esquimaux during the last week."

Sidney Budington had offended Hall by offering the Inuit passage home to Cumberland Sound on a whaling ship. Hannah and Ipiirvik had both agreed to set sail. Their time in the United States, and with Hall, could have ended here in 1863. In Hall's opinion, this was an intolerable betrayal by three people he had considered his friends, and in the case of the Inuit, his subordinates. Hall believed that the Inuit couple had entered into a verbal agreement to accompany him on his next expedition and that they still wanted to "adhere to" him.[100] The death of the couple's son, their persistent illnesses, the preparatory delays and lack of funding that threatened to keep Hall in the United

States for a second year—none of this voided the indenture in Hall's mind. To make matters worse, Budington had approached Hall's patron Henry Grinnell and had secretly asked for two hundred dollars to outfit the Inuit for their trip home.[101] On hearing of this plan, Hall rushed to New York to pick up the Inuit. He took them north of the city to Nyack for several months, far from the whaling ports during the season of Arctic departures.

The opening and closing of Arctic shipping season was one ecological and cultural marker of time that undoubtedly shaped Hannah and Ipiirvik's lives in the United States. It likely also affected their thoughts and emotions, as this was the only time of year when they could even contemplate returning north. I can't know for sure, but it seems likely that the couple wanted to return to Cumberland Sound in the summer of 1863. Hannah's attachment to her son's grave might have drawn her back to Groton again, but even a short trip back to Cumberland Sound would have given the couple a chance to reconnect with family and familiar places and foods and animals, and to pass on the news of their son's death. Spending another year away was risky for the Inuit, given their health issues, but Hall offered them no choice.

Hannah and Ipiirvik cannot have enjoyed Hall's aggressive authoritarianism, though he was typical of his time. Hall often complimented Hannah and Ipiirvik in his writings, but at the same time he treated them as servants whose needs were subservient to his own. One journal entry, written shortly before he moved the Inuit to Nyack, proclaims: "Noble, generous soul Ebierbing. He & his good hearted wife will be valuable auxiliaries in my next voyage."[102] Americans like Hall grew up within a cultural system that exalted the rights of some individuals over others. Many Americans fought against these inequalities, but African Americans, women, Indigenous peoples, and some immigrant and religious minorities experienced legal discrimination and forced subordination in the United States in a way that Inuit in Cumberland Sound generally did not. There were powerful people in Inuit society, but the boundaries of Inuit authority were more fluid, its hierarchies less absolute. "An Innuit," Hall grandly declared, "is subject to no man's control." Presumably Hall meant no control by other Inuit, since he clearly wanted to control Hannah and Ipiirvik.

It was not just American societal inequalities that differed sharply from Inuit ones, but also the forceful ways in which Hall and others displayed their power. After his first two years in the Arctic, Hall reported that he had witnessed few disagreements, and that the Inuit he saw invariably "[dealt] justly and kindly with each other," usually by keeping their distance until the quarrel was resolved, often through a mediator. In Hall's view, Inuit leaders were generally respected but had "no authority whatsoever." Hall presumably failed to recognize their authority because, unlike in America, it was rarely flaunted or expressed directly.[103]

Nineteenth-century Inuit tried to avoid open conflict. They also defined aggression and hostility much more broadly than did Qallunaat. Harsh words and even angry thoughts could sometimes kill. In the 1990s, a Cumberland Sound Elder, Etooangat Aksayuk, recounted a story about his own father Angutiqjuaq and the strength of negative emotions. The morning after Angutiqjuaq became a shaman, after being warned to be careful with his new powers, he made the mistake of thinking meanly about another man who asked to hunt at a seal hole Angutiqjuaq had found. Angutiqjuaq acquiesced, but as he was walking away, he thought, "I wanted to wait at that hole that I found, I wish he could do something else." By that night, the other man had lost his voice; he never spoke another word. Etooangat said his father regretted what he had done.[104]

Inuit adults did not generally shout, lose their tempers, or show sulkiness toward others. Those who did were feared. To Inuit who were not used to Qallunaat, angry words and harsh tones could indicate an irrational, dangerous adult who was liable to turn violent at any time. Faced with an aggressive authoritative demand, Inuit would often comply to avoid open conflict with such an immature, unpredictable, and dangerous person. There is a word, *ilira*, for the emotion that leads to compliance. It encompasses fear, but the writer Rachel Attituq Qitsualik-Tinsley says it is "not quite fear," but rather just "a feeling that it is better to yield." She noted that, "In order to understand it, you have to be able to feel it." Rosemarie Kuptana, former president of Inuit Tapiriit Kanatami, described *ilira* as "a great fear or awe, such as the awe a strong father inspires in his children or the fear of the Qallunaat previously held by Inuit." The emotional state of *ilirasuktuq* is part of the reason that Inuit so often obeyed intolerable requests from colonial authority figures. It is, indeed, impossible to know how much of Hannah and Ipiirvik's travel was truly voluntary.[105]

Charles Francis Hall seems to have lived up to the Inuit stereotype of Qallunaat who lacked control over their negative emotions. Hall probably considered it healthy, proper, and necessary to express his anger at times, but he sometimes went too far even by American standards.[106] He had some loyal friends and defenders, but he also made enemies. When the doctor on one of Hall's expeditions resigned, he cited Hall's "bad temper and his quarrelsome and jealous disposition" as the reason.[107] There is also a more serious and undisputed charge: during his second Arctic expedition in 1868, Hall assaulted and killed a whaler named Patrick Coleman.[108] Hall said that Coleman, who was unarmed, was threatening mutiny. Another whaler claimed that Hall was angry at Coleman for interviewing Inuit without his permission. Hall was remorseful after shooting Coleman, and seems to have made sincere attempts to nurse the man back to health by applying poultices of pounded hard bread to his wound, and by "bleeding" him by cutting his wrists. Coleman died after two agonizing weeks.[109] Hall was never charged with murder: the Canadian government declared it an American matter, and the American judicial system ignored it.[110] The US Congress soon approved the sum of $50,000 for Hall's third expedition.[111] Hannah and Ipiirvik had watched Coleman suffer and die. Any suspicions that Hall's verbal aggression could lead to physical violence would have been confirmed, and, for the most part, they continued to do what he wanted.

I am not suggesting that every moment the Inuit couple spent with Hall was filled with *ilira;* I do not know the scope of their feelings. After Hall's death, Ipiirvik testified that he was "a good man" and a "friend," and added he would go on a polar expedition with a man like him again. When Hall was dying and feared he was being poisoned, he would only allow Hannah or the second mate to prepare his food. He asked Ipiirvik and Hannah to sit with him. He tried to entrust Hannah with his papers, rather than any of the other Inuit or Qallunaat on board the *Polaris*.[112] From 1860 to 1871, the three spent more time with each other than with anyone else. Their relationship would have been complex and shifting, but we need to consider how, especially in the United States, it was marked by power relations and cultural differences.

Inuit found ways to disobey Hall in the Arctic, especially while out on the land, when Hall was dependent on them for survival. Ipiirvik and other Inuit guides hunted whenever conditions were good, often putting Hall's travel plans on hold. This aggravated Hall no end. The "burning indignity" of Inuit making decisions without him led Hall to exchange harsh words with a man he called Koojesse, and then to write angrily in his journal that "*I never took such insolence from any white man.*" Out on the land alone with Inuit, Hall could not command the authority he felt he deserved because of his race.[113]

Even after years of working with Hall, Ipiirvik refused to adopt a Qallunaat style of dictatorial authority. After Hall's death, Ipiirvik went on an expedition with the American explorer Frederick Schwatka, who complained that Ipiirvik would not issue commands to the other Inuit employees. When Schwatka ordered Ipiirvik to feed the dogs, Ipiirvik replied that he would discuss this issue with the other hunters later, presumably to reach a consensus on the best course of action. Schwatka became frustrated: "There is no necessity for talking it over, 'Joe;' just tell them what I say."[114] In the United States, however, surrounded only by Qallunaat, Ipiirvik and Hannah could not so easily contradict authority figures.

FIGURE 2.9. Ipiirvik. Photo from the Hannah and Joe Ebierbing Collection and reprinted courtesy of the Indian and Colonial Research Center, Incorporated, Old Mystic, CT.

Hall seems to have believed that Hannah and Ipiirvik were devoted to helping him achieve his dreams of Arctic glory and irrevocably loyal to him. In part this may have been because they rarely expressed dissent in ways that he understood, but it was also based on Hall's sense of a racial hierarchy and his position in it. He drew on the longstanding American discourses of servitude and slavery to characterize his authority over the couple. On reaching the United States, Hall referred to "the family of *Esquimaux* now under my control," language that he repeated in official contracts for their public appearances.[115] Hall often called Ipiirvik "my Joe," or "my boy Joe." Like many slave-owners who infantilized their charges to justify their domination over them, Hall's letters expressed concern over Ipiirvik's supposedly childlike innocence: he must not be allowed to drink or even to choose his own clothes. New York was a "bad place for 'Joe'" because he was naive, and "there [were] so many ways he unthinkingly might be led astray."[116] While it is true that Ipiirvik and Hannah had much to learn about the United States, Hall portrayed them as incapable of functioning as intelligent adults.

Hall's equation of the Inuit with children was explicit. They were his "dear children of the North," "almost [his] adopted children," and "almost as dear as my own little ones at home."[117] Hannah and Ipiirvik may have questioned the worth of Hall's parental devotion. He had biological children in Ohio, but he almost never visited them and does not seem to have been concerned with providing for them. When Hall quit his job and traveled to New York to seek support for his first expedition in 1860, his wife was heavily pregnant with their second child; this boy was only two months old when Hall sailed north.[118]

Hall does not seem to have noticed the irony of his Inuit "children" feeding him, clothing him, and ensuring his survival in the Arctic.[119] Even though Hall was at least fifteen years older than Hannah and Ipiirvik, on expedition he had to be provided for and often watched over like a child. Ipiirvik once remarked that American whalers—who were sent out to live with Inuit and soon returned to their ship complaining of being hungry—were "all same as small boys."[120]

Nevertheless, Hannah and Ipiirvik began referring to Hall as "Father," and in later years, their daughter would call him "Grandpa Hall."[121] This pleased Hall, to whom parental forms of address indicated deference and subservience. Hall himself sometimes referred to his wealthy patron Henry Grinnell as "father."[122] Kinship vocabulary carries different connotations for Inuit, who often address others by kinship or relationship terms rather than names. Not having a large number of relatives was unthinkable to most Inuit. Inuit sometimes "adopted" foreigners who lived with them in their dwellings and began referring to them in familial terms. Nineteenth-century Inuit likely viewed such adoptions, like all personal identities, as contingent: they may have addressed Hall as a father at some moments and not others.[123]

Hall, as "father," did not generally put the Inuit couple's needs ahead of his own desires. Preventing the couple from sailing north with Budington in 1863 was just one example. When traveling in the Arctic, he often insisted that Inuit assume the greater risks, such as going first over thin ice.[124] Even after the Inuit couple's horrible experiences at Barnum and Cutting's museums, and even when Tarralikitaq was near death, Hall wrote to the Ohio entertainer R. W. Seager, offering the Inuit family for a few weeks and asking how much he would pay—though as we have seen, he did not follow through after Tarralikitaq died.[125] For the most part, Hall seems to have been unable or unwilling to admit that Ipiirvik and Hannah's interests did not always coincide with his single-minded passion for Arctic fundraising and exploration.

The Inuit were likely motivated to acquire additional American relatives on whom they could rely. Hannah soon began referring to Sarah Budington, roughly seventeen years her senior, as her mother. In a letter to Sarah in which Hannah repeatedly referred to her as "Mother," Hannah said she was grateful to Sarah for caring for her after her son died. She told Sarah she would never forget this generosity as long as she lived. Hannah stated that Hall was kind, but in another letter her feelings were more tempered. "I like my father sometimes—father kind to me," she wrote. She lamented that her father and mother did not live together, because she could be with only one of them at a time, and she longed to see Sarah.[126] For Hannah, finding and building kinship relationships would probably have been part of making a secure home. But to Americans these letters could sound as if Hannah considered herself subordinate and unworthy of addressing other adults by their "proper" names and titles. This played into American stereotypes of Indigenous people as childlike. Although Inuit values and ideas of home benefited Hannah and Ipiirvik in many ways in the United States, they could also backfire.

A New Economy: Budington Residence, Groton, Connecticut. Sunday, October 16, 1870.

On this Sunday, the Inuit couple were once again at the Budington residence in Groton, Connecticut, where they regularly attended church. Charles Francis Hall and Sidney Budington had resolved their feud years ago. In accordance with Hall's wishes, the Inuit couple had accompanied him on a second grueling, five-year Arctic expedition, to a part of the Arctic previously unfamiliar to them. During the voyage Hannah gave birth to another son, who died in infancy, and the couple adopted a daughter in the Igloolik area. It is often reported that Hall "bought" the girl for Hannah and Ipiirvik for the price of a sled. In fact, adoption is common in Inuit society, and it is appropriate for the adoptive parents to give a gift of appreciation to the birth parents, as an assurance that they want the child to have a good life. The term for this gift is *ikauti*, which is not the same as the word for payment, *akilii*.[127] Hannah and Ipiirvik did not generally call their daughter by her name, Isigaittuq. They called her *panik*, the word for daughter, and Americans often assumed "Punny" was her name.[128]

Back in the United States, the family had recently completed yet another lecture circuit. This tour had taken them as far away as Ohio, Pennsylvania, and Washington, where they had visited the White House and met President Ulysses S. Grant. Hall was now lobbying Congress to allocate him funds for a third expedition, with the assumption that the Inuit would again accompany him.

And yet Hall's thoughts were not completely focused on the Arctic. Three days previously, Hall had written to his patron, Henry Grinnell, informing him that two acres of land, a house, and a barn were for sale for $600 near

FIGURE 2.10. Ipiirvik and Hannah's daughter, or *panik*, Isigaittuq. She was also known as Sylvia Grinnell Ebierbing. Photo from the Hannah and Joe Ebierbing Collection and reprinted courtesy of the Indian and Colonial Research Center, Incorporated, Old Mystic, CT.

the Budington residence. Hall wanted the Inuit to buy this property. He believed they would "feel greatly encouraged to go on the proposed North Pole Expedition and be faithful to it, & then return to America & settle down on their Homestead, if they can now have their just dues paid them." He asked Grinnell to send him $110 he owed him, as well as $99 owed for "[the Inuit couple's] time & assistance rendered me on my last voyage," which had ended over a year ago. These funds would serve as a down payment on the house.[129] Grinnell was the benefactor of Hall and several other Arctic explorers as well as part-owner of several Arctic whaling ships. He also acted as a somewhat autocratic bank, doling out payments and safekeeping the proceeds from the lecture tours.

The following day, a check arrived from Grinnell. It was for less than Hall had requested, but it was apparently enough.[130] Within twenty-four hours the house was purchased. The details of this financial transaction are unclear today, but it seems that the house and land were initially deeded to Ipiirvik; then Sidney Budington assumed ownership in 1871 and later transferred it back to Ipiirvik in 1875. Hannah and Ipiirvik's new neighbor, John Joseph Copp, reportedly had control over the mortgage and power of attorney on their property.[131] Copp was one of several Americans who held money for Ipiirvik and Hannah or performed financial transactions on their behalf. Most of these Qallunaat were trying to act in the couple's best interest, and some were generous with both their time and money. But it is worth discussing that by 1870, the Inuit had been working for Hall as interpreters, guides, hunters, and performers for a full decade, yet they still lacked full control over when and how they were paid, where they lived, or what kind of work they did. This hindered them from making a home in the United States on their own terms—or from returning home to Cumberland Sound. Indeed, while home ownership was a common goal for many Americans at the time, we cannot know if the couple truly wanted to use their earnings to buy a house or not.

Observers who imagined the relief the Inuit must have felt on arriving in the United States—for example, to be eating "roast beef and strawberry short cake" instead of "candles and blubber"—failed to take into account many cultural differences, not to mention facts. American food needed to be purchased rather than hunted, and wage earning was especially problematic for groups like Indigenous people, newly freed African Americans, non-English-speaking immigrants, and women.[132] Such individuals were deemed inept at handling money, and they were often forced into dependence to survive. Even Hall, in a patronage relationship with Henry Grinnell and presumably in debt to him, does not seem to have had full control over his own earnings.

Inuit likely had far more autonomy in their dealings with early Qallunaat whalers in Cumberland Sound, from whom they obtained goods like guns, ammunition, metal tools, tea, and tobacco. Although the new firearms made traditional hunting easier, without them Inuit could still kill and process seals, caribou, and other animals to meet their basic needs. They could still choose to leave the range of the whaling ships. In the United States, the Inuit could no longer slide in and out of a foreign economy. They could not live off the land; they could not build a house without buying land. They had limited direct access to the cash economy.

As a hunter in an agricultural and industrial cash-based society, Ipiirvik suffered intensely from the devaluation of his skills. He seems to have had a much more difficult time adapting to American life than did Hannah, whose expertise as a mother, cook, and seamstress fit with American gender norms and was appreciated, if not usually compensated with cash. Ipiirvik remained proud of his Arctic skills even though there were few opportunities for him to use them. In addition to building the *igluvigaq* in the yard, he fished in a local pond and carved toys, including small ivory Arctic animals, for neighborhood children. He made snow goggles for his daughter. He played a type of golf, using a rib bone for a bat and a flipper joint for the ball—quite likely a game he had learned from Scottish whalers. Yet he longed for work that would be recognized and rewarded, work that would support his family.[133]

In 1870, Ipiirvik again wanted to sail north on a whaling ship, and Hall urged Budington to "keep Joe from leaving me." He promised to find Ipiirvik a job in a machine shop. Hall noted that the man was a "*good mechanic*" and would "feel much better to have something to do."[134] A few days later, Ipiirvik dictated a letter stating that he wanted to go north because in the United States there was "nothing to do." "I want something to do," he said. "[In] my country I hunt all the time. Don't like to be lazy."[135]

Most of the jobs Ipiirvik performed in the United States were probably not what he would have considered useful work, and it is unknown whether Hall ever got him a job as a machinist. Still, he and Hannah should have earned more than enough to live on. They served Hall for years in the Arctic; they drew large audiences to his lectures; and their three weeks at Barnum's American Museum likely paid at least three hundred dollars, half the cost of their house and land.[136] When staying on a farm, Ipiirvik did so much physical labor that the landlady stopped charging Hall for the man's room and board, but Hall still refused Ipiirvik money for a trip to Connecticut.[137] Hall was chronically short of cash, but he used the excuse of the couple's "childlike" capabilities to maximize the amount of money he put toward his Arctic expeditions. His notes on expenditures in 1864, including expenses like "tobacco for Ebierbing" and five dollars "cash paid to Tookoolito to buy whatever she desires," suggest that he did not pay the Inuit a salary in the United States. On at least one occasion, he explicitly used money earmarked for the Inuit to pay off his expedition debts.[138] By controlling the couple's finances, he also made it difficult for them to abandon him. In nearly every letter Hannah wrote to Sarah, she commented on how she wanted to visit Groton. In the spring of 1864, she wrote, "I like to go see my mother. . . . I like to get dogs sleigh. That times I come see my mother."[139] If Hannah had a dog team, she could visit Sarah without needing money.

In the published account of his first expedition, Hall recounted two incidents involving Inuit and cash. In the first, Uugaq, Ipiirvik's uncle, coveted a deckhand's fiddle and offered one American cent for it. In the second, the Inuk known as "John Bull" or Johnnibo tried to pay for a one-dollar shirt with two pennies. Hall maintained that these two events showed "the simplicity of the Inuit character in matters connected with money."[140] However, Uugaq and Johnnibo presumably obtained American currency by trading with Americans. If the Inuit believed that their coins were worth far more than market value, they had likely been cheated when they traded for the money in the first place. When Ipiirvik returned to the Arctic with an understanding of the prices of whale oil and baleen and trade goods, he understood that Qallunaat were compensating Inuit unfairly, and he protested the rates of exchange.[141]

After moving into their house in Groton, Hannah and Ipiirvik and their daughter would go north once more with Hall: on the ill-fated *Polaris* expedition. Hall died on this voyage, and the Inuit family returned to Connecticut. They seem to have avoided lecture tours and public appearances after Hall's death, even though investors were eager to hire them. Hannah settled in Groton. She obtained a sewing machine and worked as a seamstress. A neighbor also remembers her sitting on the floor and chewing sealskins to turn them into fur clothing, which she either sold or gave to local people. Ipiirvik spent some time in New York in 1875 and reportedly worked as a carpenter and farmhand. He signed on with American ships most if not every summer. At least two of these voyages took him to Cumberland Sound.[142] Ipiirvik may have gone to sea because he disliked living in the United States, but I don't think we should assume it. Home is not always a single place. Many men in New England coastal communities, including career whalers, divided their lives between sea and land. Ipiirvik's house in Groton, with Hannah and his daughter, could have been a place where he felt a sense of home, but where he also felt the pull of the sea and of Cumberland Sound.

After the *Polaris* expedition, the couple's financial situation was at first precarious. Friends told Hannah to come visit them, reassured her not to worry about money, and petitioned the government for funds or equipment on the couple's behalf. Hannah wrote repeatedly to at least one government employee in Washington, DC, asking for help. In 1874, the government granted extra pay to survivors of the *Polaris* expedition, including "the two Esquimaux." One letter notes that $1,336 owed to the couple went into "their good friend Brevoort's hands." The Brevoort family had known the Inuit since their first trip to the United States, and there is no indication that the Brevoorts misused the money, but neither does the Inuit couple seem to have had full control over it. Hannah had previously drawn an allowance of $20 per month while Ipiirvik was at sea, and a few records of $20 payments in 1875 suggest that this practice continued even after the Inuit received their settlement. Indeed, the apparently frequent and generous gifts from Hall, the Brevoort and Grinnell families, and others take on a different character when one considers that these people at various times had control over the Inuit couple's earnings. At least some of these "gifts" were likely paid for out of their own money.[143]

Illness: Pleasant Valley Road, Groton, CT. Sunday, March 14, 1875.

On Sunday, March 14, 1875, Hannah was at her house in Groton. She was writing a letter to a "Mr. L. Brovent Sir," presumably Henry L. Brevoort, who controlled some of their finances. Ipiirvik had just returned to Groton from New York City because the couple's nine-year-old daughter was seriously ill. Hannah worried about the child in her letter: "I take good care of she hur . . . night and day . . . but she coff," she wrote. She ended on a more hopeful note, writing that her *panik* (daughter) was still eating and sleeping.[144] Despite Hannah's efforts, the little girl died four days later. According to a neighbor's recollection, the mother at first hugged the dead child to her, refusing to let anyone take the body away. Not only would she have grieved her daughter and her daughter's name, but she had now lost the only nearby person apart from her husband who fluently understood Inuktitut or Inuit ways.[145] The girl was buried in Starr Cemetery in Groton, next to her brother Tarralikitaq.

Accounts of Inuit abroad almost invariably describe repeated bouts of sickness, all too often fatal. Illness was the most serious environmental hazard that faced Inuit who traveled outside their homeland. The risk of dying from disease made it statistically more dangerous to be an Inuk in the United States than an American whaler in Cumberland Sound. Furthermore, while being sick in an unfamiliar place is distressing, Hannah and Ipiirvik and their children were sick in a strange place that had unfamiliar concepts of medicine and rules for survival. Inuit considered themselves bound to animals and weather and seasons, and they connected good health with maintaining proper relationships with geographical, animal, and human surroundings. Inuktitut contains many words, such as *ikullaumijuq*, that can be used to denote both weather (getting calmer) and people (recovering from illness). When questioned about this overlapping vocabulary, contemporary North Baffin Elder Aalasi Joamie considered it self-evident, stating, "No wonder! We are part of the earth, therefore the words are quite similar."[146] In the unfamiliar environment of the United States, what were Hannah and Ipiirvik's relationships with the outside world supposed to look like?

Hannah and Ipiirvik's daughter suffered from several illnesses in her short life, mostly diseases that attacked the lungs.[147] She probably died of either influenzal pneumonia or pulmonary tuberculosis. Both infections can cause bloody coughs, fevers, chills, weight loss, pallor, difficulty breathing, and chest pains. Influenzal pneumonia is a bacterial complication of an influenza virus, and it often strikes after the victim appears to be recovering. Pulmonary tuberculosis, known and dreaded in the nineteenth century as consumption, is an allergic reaction to tubercle bacilli. The body can generally destroy these bacteria, but in the process, proteins and fatty acids are released that irritate surrounding tissues. Pulmonary tuberculosis can kill children quickly, but it is most often a chronic condition that lies dormant for long periods. In some cases it never reactivates, but people under mental or physical stress—like Inuit travelers—are particularly vulnerable to relapses.

Countless Qallunaat died of pulmonary illnesses as well. Nineteenth-century cities were rife with microbes that physicians could not treat effectively. Disease was a serious risk for all Qallunaat, although it disproportionately affected poor people who lived in overcrowded housing and could not afford to eat well. As historian Bettina Bradbury writes of Montreal in this period: "Working-class families ... lived in fairly constant contact with disease, poverty, and death. The newborn children of the poor were almost as likely to die as to live. Many families were fragmented by the death of a mother or father. Many more experienced periods when one or both parents or children were sick, perhaps hovering on death."[148]

When the Inuit fell ill their first winter in New York, Hall noted that they were "afflicted the same as those of our people in many parts of our country." In 1850, nearly one-quarter of all American deaths were attributed to consumption. Tuberculosis was epidemic in Europe as well: the disease spread so easily through crowded cities that virtually all nineteenth-century Londoners and Parisians were infected with it. Hannah and Ipiirvik may have caught it on their first trip abroad to England. By the mid-nineteenth century, the disease had also spread to Cumberland Sound. Hannah and Ipiirvik had friends who died from it.[149]

When traveling with Hall in the Arctic, the couple had at least occasionally relied on *angakkuit* (shamans) to heal their bodies and reaffirm the order of the world, but they knew no shamans in the new country.[150] Hannah and Ipiirvik probably knew treatments for pulmonary illnesses that used seal or polar bear oil.[151] Even if traditional medicines often failed to cure new Qallunaat diseases in Cumberland Sound, they could still provide comfort and relief. But these ingredients were not available in the United States.

Hannah likely recognized that she needed help to understand the rules for survival in this new landscape, which had different animals, different sicknesses, different remedies, and a different God. She turned to Sarah Budington, who had previously taught visiting Inuit about her language, customs, and religion.[152] American middle-class ideals of proper societal behavior often differed from Inuit ones, but the two nineteenth-century societies both believed in a holistic connection between good health and social order. In instructing Inuit about Christian beliefs, literacy, and middle-class manners and domestic tasks, Sarah probably believed she was helping visitors stay healthy.

Sarah Budington also told Hannah the best ways she knew to specifically avoid pulmonary illnesses. The study of microorganisms was still in its infancy, and most Americans did not know that tuberculosis was contagious or what caused it. As late as 1854, some men in Connecticut dug up the body of a consumption victim and burned his heart to stop him from infecting others from beyond the grave.[153] Most American doctors would have scoffed at this, but neither did they believe tuberculosis was contracted through airborne contagion, since relatively few people showed symptoms immediately after contracting the disease. The major causes were thought to be heredity, poverty, dirt and dampness, the excessive consumption of alcohol, and cramped and uncomfortable working conditions. Most of these stresses could indeed have sparked a reactivation.[154] In keeping with these beliefs, Sarah advised Hannah to "never sit down on the ground and never get [her] feet wet and never sit down in the wind." Hannah took this advice seriously, confessing to Sarah in a letter that she sat down on the ground one hot summer day. Soon afterward, she caught a cold and regretted not having heeded Sarah's advice.[155]

Some American medical advice would not have been surprising to Hannah. Both Qallunaat and Inuit societies stressed the importance of being active, of eating wholesome and simply prepared foods, of wearing appropriate clothing, and of breathing open air. Both knew that negative emotions could lead to illness. In other ways, however, they differed. American health experts of the period recommended washing daily, linking lice to filth and poverty. But at least some Inuit believed that the relationship between lice and humans was symbiotic rather than parasitical: in draining blood, they cleansed the body of illness.[156] It seems that in the United States, Hannah adopted American ideas of propriety and healthfulness. Hall commented that she was "pleasing and refined in her style and manners," implying that in his presence, she followed his rules.[157] Indeed, by the end of her time in the United States, Hannah had acquired three main characteristics that Americans associated with "civilization": private property, Christianity, and fluency in English.[158] This probably allowed her to obtain acceptance, respect, and aid from Americans who might otherwise have shunned her.

Both Hannah and Ipiirvik expressed gratitude for Sarah's help and advice. Hannah would not have been the first woman Sarah had nursed or helped through grief: living near a major whaling port, women took care of each other while their husbands were away or after they were lost at sea. Ipiirvik reportedly commented that Sarah was a good friend to Hannah because she "come to see her—help her—tell her what to do."[159] When Hannah was bedridden in New York City in 1864, she wrote to Sarah, recalling her past help. "When I was sick you take care of me," she remembered. "So good to me, you take care of me many weeks."[160] Presumably Hannah and Ipiirvik's gratitude was sincere. However, sickness has long influenced power relations between Qallunaat and Inuit; expressing gratitude for health care can also be about keeping on the right side of people who have some control over your life.[161]

Hannah and Ipiirvik's gratefulness does not imply that they accepted all of Sarah's advice uncritically, or that it replaced their old ways of thinking. It seems that they tried to incorporate Inuit methods of healing along with American cough syrups and doctors' visits.[162] Contemporary North Baffin Inuit Elders hold that sick people should be given whatever food they most wish for.[163] In her letters, Hannah wrote of her husband trying to do this for her, by obtaining wild birds, like partridge, that she requested while ill in New York City. When Hannah's daughter was ill, she reportedly craved raw meat, which Hannah bought for her.[164] It seems that Hannah also sought substitutes for seal oil, which Inuit fed to people with lung infections and sometimes rubbed directly on the skin. She bought castor oil at the local store when her *panik* was sick.[165] Still, she and her husband sometimes longed for their known remedies. Ipiirvik spoke of wanting seal to cure his cough.[166] Perhaps even more importantly, the couple were no longer surrounded by the Inuktitut words, Inuit spirits, and Inuit companions that they would have considered crucial to maintaining good health back home.

Americans noticed that Inuit seemed susceptible to serious illnesses. After the death of Tarralikitaq in 1863, one article pointed out that "hardly one of the Esquimaux brought either to England or this country escapes some very serious sickness, if not death."[167] Yet most Americans were so persuaded by the notion of the Arctic as a harsh environment that they could not imagine the northeastern United States as less healthy. Qallunaat reports did not describe the temperate places where Inuit died as inherently "unhealthy" or "sickly," even though Qallunaat often applied these terms to environments where they themselves were susceptible to illness. They instead came up with alternative explanations for Inuit deaths. Some blamed earlier privations in the Arctic, including the horrifying months spent on an ice floe during Hall's last expedition. Others argued that humans—or at least non-Qallunaat—could only flourish in their home climate. This idea of a "natural" home for different peoples was a common Qallunaat belief, and many feared for their own health when moving into new territories. Remember that before Hall went north to live among Inuit for the first time, he had been anxious about whether "White men" could survive among Inuit.[168]

Inuit, the Qallunaat reasoning went, were a strain of humanity somehow better suited to cold than to heat, to raw over cooked. Newspaper articles and other accounts repeatedly stated, implicitly or explicitly, that Inuit visitors were sick because they were not meant to eat American food or live in temperate regions. The summer weather destroyed them, the dense ocean fog killed them, the "wet and changeable" American climate "produced" disease in Inuit.[169] The last writer was presumably unaware that Cumberland Sound summer weather was nothing if not wet and changeable. Hall seemed to think he was gambling with fate every time he brought the Inuit south. "Take the Esquimaux away from the Arctic regions," he wrote, "and they would soon cease to exist from the face of the earth. The bounds of their habitations are fixed by the Eternal, and no one can change them."[170]

Inuit must have had less biological immunity to tuberculosis, which had existed in much of North America for millennia, but not in Inuit territories. Nevertheless, other causes like poverty, malnutrition, poor access to health care, and the stress or brutality of colonial encounters often contributed to high mortality rates among Indigenous populations.[171] The immersion of Hannah, Ipiirvik, their children, and other Inuit visitors in a foreign landscape and culture almost certainly weakened their immune systems. The emotional stress from dealing with public scrutiny, constant travel, aggressive authority figures, and a new and discriminatory economy would have compounded any lack of immunity they had.

On Sunday, December 31, 1876, Hannah died at home in Groton of "disease of the lungs," almost certainly tuberculosis.[172] She was thirty-eight years old. Ipiirvik was at sea until a month before her death, on a voyage with the US Fish Commission, so it was probably Sarah Budington who tended to Hannah as she "[suffered] greatly from consumption." Hannah had weakened following the death of her daughter the previous year. Some of her last words were reported to be, "Come, Lord Jesus, and take thy poor creature home!"[173] This may well be true—Hannah had learned to read the Bible—but it may also be Victorian sentimental invention. Several newspapers published articles about her death, recounting the Arctic exploits that had made her a minor celebrity. Much less was said about her life in Groton. An immigrant seamstress succumbing to tuberculosis was hardly remarkable in 1876.

*

It was no easy feat for Hannah and Ipiirvik to live abroad for as long as they did. They likely did not always feel they had a choice, but in 1873, after the death of Charles Francis Hall, they reportedly decided to stay in the United States. After their six-month drift on an ice floe, they were brought to Washington to testify in the *Polaris* expedition hearings along with all the other survivors. Hans Hendrik, the other Inuk father from the floe, told government officials that he absolutely did not want to remain in the United States. It was too hot; his children were sick; he wanted "to go home right off."[174] The US government returned him and his family to Greenland. Hannah and Ipiirvik chose their house in Groton. Why? Did they both want to live there? And did they consider it a permanent move?

Soon afterward, Ipiirvik stopped in Cumberland Sound on a voyage to help rescue the remaining *Polaris* crew. It was his first opportunity to return since sailing south with Charles Francis Hall eleven years earlier. We have no records of his two-week visit, but he brought his half-brother, likely named Ittuluk, back to Groton with him. Ittuluk spent the winter with Hannah and Ipiirvik. He went home to Cumberland Sound on a whaling ship in the spring of 1874 and died shortly thereafter. Before he passed away, he dictated a letter to Hannah and Ipiirvik, saying he planned to stay in Cumberland Sound for two years and then visit Groton again. He longed to see them both. He asked for some cartridges for his gun, and offered to send fox furs and seal skins and caribou clothing. He promised to send more letters, and asked Hannah and Ipiirvik to write to him care of the Qallunaaq whaler John Spicer's ship.[175]

Hannah seems to have found a home she could accept in Groton. The records of her time in the United States are sad, but also not nearly complete. She wrote most of her letters to Sarah when she was grieving and lonely, and Qallunaat sources most often reported on the Inuit when they were in public forums or suffering from illnesses. There were long periods of uneventful silence in Groton, which likely contained unrecorded visits with American friends, feasts of Qallunaat foods Hannah had known since childhood, relaxed periods of teaching her children both Inuit and Qallunaat ways and watching them grow. There is no record of Hannah trying to return to Cumberland Sound after 1863, and by the end of her life she had spent so many years abroad that it is doubtful she would have felt fully at home there.

Following Hannah's death, Ipiirvik remained in Groton for another year and a half, where he was seen weeding the grass on his wife's grave.[176] However, his sense of home there may have been contingent on his family's presence. According to one report, he had tried to convince Hannah to return to Cumberland Sound after their daughter's death.[177] In June 1878, he sailed on another American expedition, to a part of the Arctic that he had earlier visited with Hall. Before leaving, he voiced distrust of the Inuit who lived there, but apparently said that with Hannah and his daughter gone, he no longer cared if he put himself in danger.[178] He married a woman recorded as Neepshark, and never again returned to the United States.[179]

Some Inuit formed strong attachments to people and places in the United States. But like American whalers in Cumberland Sound, very few Inuit chose to overwinter more than once. Ittuluk, who stated a desire to return a second time, had the unique opportunity to live with Inuit relatives. Hannah and Ipiirvik had worked hard for over a decade to become more autonomous and respected, more in control of their own finances and mobility, more knowledgeable about their new surroundings, and to create a space where they and their relatives could be comfortable. Yet the United States remained a difficult and dangerous place, one that Inuit were both interested in and scared away from. On Ipiirvik's final expedition, a young family reportedly wanted to return to America with the explorers. Ipiirvik spoke "of the great mortality attending those of his people from Cumberland Sound who had gone to England and America," and several Elders in the community decided that the family should not travel after all.[180]

3

Americans and Inuit in the High Arctic

On January 13, 1881, Lieutenant Frederick Kislingbury penned a letter and marked it "Personal!" He had just lost his second wife to typhoid fever in Montana Territory. His first wife had died of illness less than three years previously. The letter's recipient, First Lieutenant Adolphus Greely, was heading up an American expedition to the Arctic.[1] Kislingbury pleaded to go:

> My friend, you will not wonder if I speak of grief and sorrow . . . With you, up there in the cold north, I can find relief. It will be like leaving a world that has been so cruel to me. I can find up there hard work & plenty of it. Overland trips through snow and ice and the kind of exposure that will do me good. Ah! My friend, the future looked very dark to me & your good letter comes as a boon. It awakens me from a fit almost of despondency. It seemed as though all sources of joy and even of sorrow were drying up in me. But now the future looks brighter. The separation from my children will be as nothing compared to the prospect there will always be of having been with those who may accomplish some great and lasting good to mankind . . . even should nothing be accomplished my children will love me better when I return & will be proud of the father who dared to brave the dangers, the depressing influences, in fact, everything we imagine and have read about of a sojourn in the Arctic regions. . . . If I go, I feel that I shall come back a new man.

Kislingbury signed on as second-in-command, leaving his sons with relatives. He wanted a stark break from his surroundings and state of mind; he wanted to go somewhere that was the opposite of home. Home is not always safe and comfortable; it can also be a hollow of pain, something to escape and leave behind. For Kislingbury, a trip to the Arctic seemed "a wonderful chance for me to wear out my second terrible sorrow."[2]

Six months later, Kislingbury was on an expedition ship anchored off the coast of Greenland. Greely had received far more applications from American soldiers than he could accept. But now, when they were stopped at Upernavik to hire Kalaallit (West Greenlandic Inuit) employees, they could not find anyone ready to sign up. The local Danish inspector recommended two men from a nearby settlement. These men joined the expedition without open protest, although it is unclear if they wanted to go.[3] The ship headed north for Lady Franklin Bay, a place that was home to no one on board. It was about three thousand miles north of Washington, DC, a thousand miles north of Cumberland Sound, and six hundred miles north of Upernavik. The expedition's stated purpose was recording scientific observations concurrently with other stations around the Arctic, in the cooperative spirit of the first International Polar Year. Greely would also launch sledge trips toward the North Pole to claim the Qallunaat "Furthest North" record for the United States.[4]

By the time the ship reached its destination, Adolphus Greely and Frederick Kislingbury were already at odds. Following one final quarrel over the timing of breakfast, Kislingbury requested to be relieved from the expedition. Greely agreed, but the departing ship got under way before Kislingbury could reach it. He was left on shore: no longer an official expedition member, in debt, and ultimately lost.[5]

Three years later, this small crew would make headlines around the world. After resupply ships failed for two successive years to reach them, the twenty-five men began a desperate retreat southward. They encountered no ships or people. Seven survivors were rescued in June 1884. The others had all died, mostly from starvation. Those who returned did indeed come back as "new men," but broken and reshaped in ways they could not have foreseen. In Kislingbury's last moments, he dreamt of American food and weather and his children, and chastised himself for allowing his thoughts to "wander too much towards home."[6]

Polar expeditions, particularly disastrous ones, came to dominate Qallunaat conceptions of the Arctic, even though they represented only a very small slice of human experience in Inuit territories. In this chapter, I consider what is left out when Qallunaat fixate on these kinds of horror stories, which often occurred in places that were remote even to Inuit. I also reflect on what we can learn about home from the thousands of pages of records of this expedition, by focusing on the experiences of two low-ranking members who perished: Jens Edvard Angutisiak from West Greenland and Sergeant Hampden Sidney Gardiner from Philadelphia. Angutisiak and Gardiner did not have the same worries, fears, and dreams when they looked out the window of the expedition cabin. Nor did they have the same opportunities to try to re-create some

FIGURE 3.1. Qallunaat members of the Lady Franklin Bay Expedition. A few men's faces were added over the heads of initial expedition members who dropped out for various reasons. Greely, *Three Years*, vol. 1. Image courtesy of the University of Calgary Special Collections.

sense of home and comfort on Ellesmere Island. They did not live and die on equal terms. Gardiner and Angutisiak's recorded opinions and actions make clear that there is no universal way to react to an unfamiliar environment, and they show how much human relationships and power dynamics affect our experience of place.

The High Arctic environment is as vital to this story as the dreams and anxieties that people spread over it. Ellesmere Island was and is part of Inuit territory. Many Inuit who had never been there knew about it, and groups from northern Greenland and as far south as Cumberland Sound traveled to the island in the nineteenth century. Human presence in the region goes back at least four thousand years, waxing and waning due to climatic shifts and other factors. At certain times and places there has been intensive seasonal or year-round use. As historian Lyle Dick explains, Inuit and Qallunaat lived in the region at other times without mishap, so the place itself is not responsible for the death of nineteen men. The Lady Franklin Bay expedition failed for a variety of reasons, mostly human ones, including a lack of local knowledge, poor ice conditions, overconfidence in western technology, a hierarchical military structure, and individual foibles of certain personnel.[7] But still it was a place far more isolated than New England or Cumberland Sound. While the Lady Franklin Bay expedition members saw archaeological evidence of human occupation, they encountered no other humans during their three years away.

The records suggest none of the members felt at home. They were indefinitely cut off from the places and people they knew. For the Kalaallit employees from West Greenland, the animals and landscapes were somewhat legible, and they would have heard stories about Ellesmere Island, but they had never been there. They were also immersed in a foreign American military culture. As for the soldiers, they had left a nation that seemed to be speeding up, pushing out, rushing inexorably forward. In 1881, they moved mostly by train from their scattered posts to Washington, DC, and then steamed north to the edge of their known world. Their initial journey embodied the modern annihilation of space by time. Yet when the Americans disembarked, they found that time was annihilated—or at least skewed—by the vastness of space.

The High Arctic landscape helped to shape all the men's moods, emotions, behavior, and senses of identity. Familiar markers of time and distance were absent or distorted. These included friends and family, place names, stars, wind patterns, settlements, vegetation, landforms, buildings, and the familiar circuit of the sun. The sun's movements stood out. Northern latitudes are characterized by summer light and winter darkness, with the Arctic Circle roughly marking the line where the sun neither rises on the winter solstice nor sets on the summer one. On northern Ellesmere Island, the sun circled the sky all summer, but when it dipped below the horizon in October, it did not rise again until February. The Americans fixated on its reappearances, and these solar patterns were reportedly distressing for the Kalaallit employees too, since they came from farther south and were not accustomed to such long periods of darkness and light. This chapter passes through four solar occurrences and discusses events related to them: the first disappearance of the sun in October 1881, the December 1881 winter solstice, the final reappearance of the sun in February 1884, and the June 1884 summer solstice when the survivors were rescued. The first two sections examine what the expedition members, especially Angutisiak and Gardiner, feared during their first winter in the High Arctic, and how this affected their behavior. The final two sections deal with a time when some of their fears were realized and they were starving to death.

The Disappearance of the Sun in 1881

The men arrived at Fort Conger in Lady Franklin Bay in August 1881, when there was still constant daylight. They constructed a cabin out of lumber they had brought with them, showing little knowledge of or concessions to Inuit architectural styles.[8] They then settled into their routines of scientific mea-

FIGURE 3.2. Constructing the expedition cabin at Lady Franklin Bay, August 1881. LC-USZ62-136209, Library of Congress.

surements, daily chores, and occasional sledge journeys. The days quickly grew shorter.

On October 14, Greely reminded the men that this was the last day they would see the sun that year. They rushed outside to witness the sun rise and set. A thin glowing crescent slid over the horizon, "sprinkling the ice and snow with silver and crystals, and then [sinking] back in a beautiful glow of warm rosy colors." It was a dazzling sight but left, one man said, "a twinge of sadness . . . to think this was the last view we had of the sun for 136 days!"[9] In the weeks that followed, there were decreasing periods of ambient light, and within days, bright stars could be seen at noon. An expedition member reflected that, "With [the sun's] departure, a cloud seemed suddenly to have been thrown over our lives; the loud merry voices were hushed, and each appeared lost in his own reflections."[10] The darkness ushered in a period of fear and melancholy.

The soldier who perhaps best recorded the shock of that first winter, and the unsettling sense of not being at home, was Sergeant Hampden Sidney Gardiner of Philadelphia. Sergeant Gardiner had attended public school before being apprenticed to a scientific instrument maker. He then enlisted in the army and joined the Signal Corps. He married just before leaving for the Arctic at age twenty-two.[11] In Lady Franklin Bay, Gardiner worked hard as a meteorological and tidal observer, fixed instruments, collected fossils and other specimens, did his share of tedious chores, and produced skilled drawings.[12] At nearly five feet nine inches, he was considered tall and of strong build. Yet he proved to be one of the sicklier men. Before the starvation period, Gardiner

FIGURE 3.3. Sergeant Hampden Sidney Gardiner. National Archives (US) photo no. 200S-LFB-111.

was treated for or diagnosed with severe headaches, a painful carbuncle on his finger, "neuralgia of the jawbones," and "anoemia and derangement of the digestive organs."[13] He seems to have been a well-liked, quieter presence on the expedition.

Gardiner was active and athletic when he first arrived in Lady Franklin Bay. He went on various short exploratory trips, hauled ice for drinking water, and took bracing walks with other men in his spare time.[14] Everything changed in November 1881. After the sun had set for the winter and all light from it had vanished, Gardiner was making tidal observations in the early afternoon when he slipped and fell on the dark icy pathway leading from the cabin to the tide-gauge. He broke his left leg just above the ankle. He called out for help, but no one heard him. Gardiner eventually succeeded in crawling to the cabin door. He lay in the snow, exhausted and crying out, until the cook found him.[15] He was off duty for several months while his leg slowly healed.

In early November, weeks before his injury, Gardiner had already been concerned about the winter ahead. He wrote: "Our day is now as one long night and for the last month we have been burning lights all through the 24 hours. We have all long since commenced to count the weeks and months to the time when old "Sol," will reappear to us in the spring, I do not think we have had a chance yet to feel the depression of spirits arising from being shrouded in continual darkness."[16] Gardiner's notes reflected a common worry: what would the darkness do to the party? Men had perished or disappeared or gone crazy on Arctic expeditions. Some people at the time believed that darkness, cold, and damp could cause scurvy, even with daily lime-juice rations.[17] Several of the expedition members had read works of earlier American explorers like Elisha Kent Kane and Isaac Israel Hayes. Kane claimed that the most trying thing in Arctic exploration was "this constant and oppressing gloom, this unvaried darkness."[18] Hayes described the dark season this way: "For say what you will, talk as you will of pluck, and manly resolution, and mental resources . . . this Arctic night is a severe ordeal. Physically one can get through it well enough . . . but it is, nevertheless, a severe trial both to the moral and the intellectual faculties. . . . The dark and drear solitude oppresses the understanding; the desolation which everywhere reigns haunts the imagination; the silence—dark, dreary, and profound—becomes a terror."[19] Some polar explorers likely exaggerated their winter sufferings. Most readers would never experience a polar night, which left narrators freer to embroider its terrible effects. But the expedition members were left to wonder how they would react. None of them had ever been this far north, and only the two Kalaallit and the civilian doctor had previously lived through a winter north of the Arctic Circle. By early February, Lieutenant Lockwood noted that everyone had become very pale, and hoped that "the darkness has had no worse effect."[20]

Gardiner was less hopeful. In February 1882, soon after returning to work, he wrote that he had passed through a "strange experience" of four months without the sun, and that "this continual darkness is terrible in its effects on man and beast."[21] Sergeant David Brainard agreed, stating that after a few weeks of darkness, the "effect . . . on the men is very apparent," with most of them either depressed or irritable.[22] Although people tried to hide their low spirits and put on a show of cheerfulness, Brainard believed the lack of sunlight affected everyone. Greely later called it "darkness so continuous and intense that the unsettled mind is driven to wonder whether the ordinary course of nature will bring back the sun." Although he may have been overstating his own fears, he claimed that the Arctic winter "almost [unsettled] the reason"—no small concern for a nineteenth-century scientist.[23]

The lack of sunlight was almost certainly not the sole cause of the men's depression and irritability. Our sense of well-being is always based on—among other things—the larger environment, our shelter, and the relationships playing out within these places. Expedition policy confined the men in close quarters throughout the winter. Hunting and exploratory sledge trips ceased. Everyone stayed indoors except for scientific observations, short walks, and chores such as ice-cutting.[24] The enlisted men lived and slept in a single room,

FIGURE 3.4. Lieutenant Greely's corner; he had a personal space in the cabin that could be closed off by curtains. There are no images of the enlisted men's bunks. The back of the image notes that it was taken "the day the sun left," October 14, 1881. LC-USZ62-136204, Library of Congress.

four to a double-tiered bunk. According to Lieutenant Lockwood—who as an officer had his own bed in a private corner—such conditions exacerbated personality conflicts. The cabin was "tolerable" during the summer months but "to pass an Arctic night under such circumstances must be experienced to be described."[25] The following year, as soon as the days became warm enough, one man moved bedding and a desk into a tent outside, where he could "sleep in purer air and be by [him]self."[26]

The claustrophobic winter living arrangement was particularly trying for reserved and quiet individuals like Gardiner. In his diary he records his annoyance at the revelry that often continued well past the official bedtime of 11 p.m. He had grown up in a large family and then boarded in the home of another, so he was presumably used to a lack of privacy indoors. But now, especially with his broken leg, he could barely leave the house. The repetitive nature of the winter tasks also grated on him. "Our life is very monotonous," wrote Gardiner, "usually the same thing over and over, day after day. Most every day some allusion is made to the ship coming next year and as to who will be the lucky men to return in her."[27] Only a few months into the expedition, he and others were already disillusioned with their Arctic adventure and hoped to flee this place as soon as possible. Their ship never came. The men had been dropped off in a year of particularly clear seas, and the ice pack would prove too thick for ships to enter Lady Franklin Bay in the succeeding two summers.

It was not just the darkness, the cramped quarters, and the monotony that bothered the men, but also the silence and the vastness of the wintry world outside their cabin door. All the Qallunaat on this expedition had volunteered for it; they deliberately sought out contact with what was to them a mythical place. Now that they were there, they couldn't ignore this seemingly endless landscape with its endless weather. Humans constantly interpret and shape environments, which then become part of ourselves. If we are comfortable or excited to be in a place, this sense of the outside world ebbing and flowing into us can feel calming or even transcendent; but for the Lady Franklin Bay expedition members, it was more often experienced as pain, fear, and trauma. The landscape seeped into the expedition members' drafty house during the long hours they spent inside. It blew through their inadequate clothing whenever they had to work outside. It got into their minds.

Gardiner and other Americans were at first in awe of their surroundings. One sergeant described the aurora borealis as "something wonderful to behold, [possessing] a fascination which leaves one speechless with surprise and wonder—impossible to describe." "The beauty of the sky was incomparable," he noted, and the ice "glistened with a cold splendor like the illumination of a thousand fairy lamps."[28] Yet these sights soon wore thin—by December 1881, another man noted that auroral displays occurred almost daily and had "ceased to be objects of attraction."[29] Even as the expedition photographer commented on the beauty of ice floes in 1882, he nicknamed one with particularly large icicles "an Arctic prison."[30] At least in their writings, the soldiers seem to have quickly found their surroundings more disturbing and tedious than thrilling, even while they were still well fed and expected to return safely to the United States.

Sergeant Gardiner's most striking critique of the Arctic landscape came on a beautiful day in March 1882. The sun had recently reappeared, and Gardiner gazed up into a stunning azure sky with cumulus clouds "tinted by the rays of the sun to all colors." After describing the magnificence of the scene, he moved onto the somber thoughts it had inspired, writing:

> The air calm and still as death, not a thing stirring with life. All this makes the scene one of terrible sublimity. When out alone and with no living thing in sight I have often stood quite still and listened to try and catch some sound, but not the faintest was to be heard, and by and by having lost my self in thoughts and wonder at "God's" work, which in these latitudes is forced upon one's

mind at every step, I have started suddenly as though affrighted, and hollered with all the power of my lungs to try and break the awful stillness and quiet which is so oppressive. But how vain are my efforts; my voice dies away quickly and at its greatest effort sounds like some infinitely diminutive creature lost in space. But enough of this if I allow my self to continue in this strain any longer I will certainly get a fit of blues, so back to plodding work, which is the only available preventive in these regions. My impressions of my life in the North are lasting and I can write and think of them when I return to the states, where they will not be so likely to bring on unpleasant thoughts and feelings.[31]

For Gardiner—a man so devoutly religious that he later brought his Bible among the eight pounds of gear he was allowed on the retreat—putting "God" in quotation marks was damning.[32] Lady Franklin Bay, with its almost inconceivable immensity and apparent permanence and silence, impressed on Gardiner that he was small and weak and alone, an "infinitely diminutive creature lost in space." It was unlike anywhere he or the other American soldiers had experienced before. Even those who had applied from western posts were linked by steel and horses to the clamor and cacophony of settler cities. Although they may have had ambiguous feelings about all the changes, they were agents of the state helping to move that world westward, some of them through armed conflict. All the soldiers benefitted from ever-faster mail and shipping services that connected them to commercial and military networks, and to loved ones or their childhood homes. In Lady Franklin Bay they were cut off. The perceived desolation seeped into Gardiner, and it reinforced his loneliness and his fears of personal insignificance. Like Kislingbury, he turned to hard, "plodding" outdoor work to break an unwelcome state of mind.

Gardiner characterized the High Arctic landscape as a terrifying void, lifeless and noiseless. Other American expedition members concurred, calling the region the "land of desolation"[33] and commenting on its silence. Commander Greely noted that even his quiet rural home in New Hampshire was enlivened by the constant "indistinct hum of insect life." In the Arctic, the snow absorbed sound, and he heard nothing except the grinding of the ice.[34] It is not surprising that the lack of noise disturbed the Americans at Fort Conger. For urbanized westerners, sound denotes vitality and movement through time; its absence is death, monotony, and perpetuity.[35] Wealth and progress are associated with high levels of ambient noise. At Lady Franklin Bay, the two Kalaallit employees presumably missed some sounds from their homeland: their language, laughing children and family members, women scraping skins, the calls of some Greenlandic animals and birds. Yet they probably would have been far more uncomfortable with nineteenth-century urban soundscapes than with the relative silence of Lady Franklin Bay.

In philosopher Gaston Bachelard's 1948 work *La terre et les rêveries de la volonté,* he analyzed experiences where massive scenery concords with vast, transcendent, dreamlike thoughts. For Bachelard the word "vast" had positive connotations, and immense landscapes opened the mind to new dimensions of beauty, power, peace, perfection, and the sacred.[36] Other theorists have echoed Bachelard's enchantment with the vast. In cultural theorist Jean Baudrillard's road trip through the American Southwest in the 1980s, he found the desert landscape "alive with a magical presence." It was a place "where the air is so pure that the influence of the stars descends direct from the constellations."[37] For Baudrillard, "space there is the very form of thought."[38] But vast landscapes such as deserts or the Arctic can remain mythical only at a safe distance, when witnessed out the window of a climate-controlled space, or described from the comfort of an office chair, or imagined by the grieving Kislingbury before he signed on. For the Americans on the Lady Franklin Bay expedition, to spend years in such a starkly foreign place was to become all too aware of the basic needs of life, of their inability to supply them, and of attachment to faraway homes. As Gardiner put it, this environment "forced [itself] upon one's mind at every step." If space is the very form of thought, thought is also the very form of space.

In the end, however, the Americans emerged relatively unscathed from their first Arctic winter. Their spirits mended along with Gardiner's broken leg, and they remained relatively healthy. They ate well, likely better than many of their families at home. Their diet was a varied mix of imported food mixed with some local meat, especially musk ox. No one developed scurvy, although the doctor kept his eye on a few weaker members. Sergeant Brainard wrote encouragingly to his mother that although the darkness had subdued some exuberant spirits, at least to his knowledge "we were never visited by melancholy thoughts or the feeling of despondency in the exaggerated manner usually described by those who had preceded us."[39]

On February 28, 1882, many expedition members hiked out to a nearby island to watch the sun rise for the first time and "roll along the horizon like a huge ball of flame." They let out a spontaneous collective shout and one man commented, "A great load appeared to be lifted from our hearts."[40] Several others remarked that they never again intended to let the sun set on them for longer than a night.[41] Gardiner himself did not see the sun until it reached the cabin on March 5, having worked all night on February 27 and slept through the next day. He wrote, "May it never go down again . . . [the sun's return] affected me very much. I fully realized how much joy its light is capable of bringing anyone. In future the nearer I can get to it, the more I can feel of its effects the happier I will be."[42]

Winter Solstice 1881

The winter solstice occurred on December 21, 1881, a fact not lost on the men, who welcomed the passing of the shortest day of the year.[43] The expedition newspaper, the *Arctic Moon,* stated that it had "made complete arrangements to have the Sun interviewed on his return to the country the latter part of February." The paper was so called, the editors explained, because the moon gave them light and solace in the darkness, and as a feminized celestial body, it reminded them of the women they missed at home. Inuit also associate the moon with winter, but reverse the gender: the moon is the brother and the sun the sister.[44]

For Christmas a few days later, the soldiers decorated their quarters with items they had on hand: American flags, crossed swords and guns, and homemade paper flowers. They also lit the room more brightly than usual, and the decorations served to cover up the coal soot that coated the walls. On Christmas Eve presents were distributed. Some men had brought gifts from home, and everyone received small items from donors, primarily pipes and tobacco. Greely wrote that some of the lonelier men cried when they received these gifts from strangers.[45] On Christmas Day the men held a service and prayed for all the "dear ones at home." The Christmas dinner menu, which had eight courses, was a mix of packed treats like Mrs. Greely's plum pudding and exotic dried fruits, and local food such as eider duck and musk-ox tongue. It was by anyone's standard an elaborate and luxurious meal. "Every one is happy," recorded Private Schneider.[46] Yet there is considerable evidence that the two Kalaallit employees, who also would have celebrated Christmas back home in Greenland, were not happy that December.

A few weeks previously, Jens Edvard Angutisiak had disappeared. December 13 had been one of the darkest days of the winter: dim and overcast with no moon or stars, with "a dense fog covering & enveloping everything."[47] Early that morning, Angutisiak had walked away from the soft lantern glow of the expedition cabin without taking any source of heat or light. He left without eating breakfast, and without any food, weapons, or ammunition. The temperature was about −28°F (−33°C), but he left without his mittens and warmest clothing.[48]

Like Sergeant Gardiner and so many other expedition members, Angutisiak was homesick on Ellesmere Island. His settlement of Kangersuatsiaq—then known officially as Prøven—was an island community of around a hundred inhabitants, nearly all of whom were Kalaallit living in sod houses heated with seal-oil lamps.[49] In the nineteenth century, there was commercial beluga whaling in the area, and local people depended on many of the same Arctic

FIGURE 3.5. Angutisiak and Dr. Octave Pavy skinning a seal at Fort Conger. LC-USZ62-136197, Library of Congress.

mammals and birds found in Cumberland Sound.[50] When the Lady Franklin Bay expedition hired Angutisiak in the summer of 1881, he was thirty-eight years old. He had four children, and his wife, Anna Maria, was two months pregnant.[51] He would later speak to Commander Greely about how much he missed his "pickaninnies."[52]

It is unclear how Jens Edvard Angutisiak and the other hunter, Thorleif Frederik Christiansen, felt about joining the expedition. Greely wrote that they had "only been obtained through strenuous exertions."[53] They would have had some idea of what expedition work entailed. Another man in their small community, Hans Hendrik, had worked for four Arctic explorers; he and his family had spent six months on the ice floe with Hannah and Ipiirvik. Perhaps Angutisiak and Christiansen were willing to risk dangers, inconveniences, and indignities for a promised salary and future opportunities. Almost certainly they were reluctant to refuse the Danish inspector and governor; the former had recommended hiring them specifically and the latter came with the Americans to pick them up. Inuit often felt compelled to comply with requests from powerful authority figures, fearing repercussions for themselves and their families if they did not. In his memoirs, Hans Hendrik recorded a lack of enthusiasm in joining all four of his expeditions. On the last one, in

1876, the local Danish trader simply told him he was to go, and Hans "reluctantly agreed" and joined the foreigners, giving the appearance to them that he welcomed the employment.[54]

Regardless of Angutisiak and Christiansen's reasons for joining up, they struggled to understand and live within their new surroundings. Inuit generally find their home solar patterns to be unextraordinary. Those who live in areas far enough north to have lengthy periods of winter darkness still perceive circadian days and nights; they do not linguistically speak of "one long night" as outside observers often do. Two contemporary Inuktitut dictionaries make no reference at all to light and darkness in their definitions of the seasons.[55] The moon remains important through the winter, and from the movements of the stars Inuit know when the sun will return in their home region.[56] But Angutisiak was not at home in Lady Franklin Bay. Long months of sunlight followed by only moon- and starlight were new to him, too. We have little direct record of his reaction, but other Inuit have reported feeling intense dislocation and homesickness in the High Arctic. Hans Hendrik recalled that the first time he encountered the constant darkness on an expedition, he "fell a weeping, I never in my life saw such darkness at noontime. As the darkness continued for three months, I really believed we should have no daylight more."[57]

Angutisiak was probably not overwhelmed by the vastness of the land like Gardiner and many of the soldiers were. Growing up in Greenland, he would have already learned to feel small in the world. Lady Franklin Bay likely seemed empty to him in certain ways: he was forced to navigate the area without known landmarks or travel routes, and there were no Inuit communities there. But this did not mean it was uninhabited. In Inuit traditions, the land and sea are home to supernatural beings. Angutisiak had been baptized a Christian, and he lived in the household of his father-in-law, the local lay preacher. But he almost certainly retained Inuit beliefs as well as Christian ones. He was a hunter and spent a great deal of time out on the land in a place where shamanistic and Christian entities coexisted.[58]

In West Greenland, as in most Inuit homelands, life centered on the coastlines and marine animals, although people traveled inland seasonally to hunt caribou. These inland regions were the permanent home of a range of dangerous beings, including ghosts, giants, dwarves, monster animals, *eqqillit* or dog-people, and *qivittut*. The *qivittoq* (singular of *qivittut*) is the creature most obviously relevant to Angutisiak's disappearance. *Qivittoq* stories have been well documented in Angutisiak's home community of Kangersuatsiaq, both in the nineteenth century and today.

Qivittut have human origins. They begin as people who flee society due to mistreatment. Common reasons for going *qivittoq* included not being allowed to marry; losing face; being physically, sexually, or mentally abused; or being scolded harshly and unfairly. Running away from an Inuit community was an act of desperation. Angutisiak's society was strongly communal, and the *qivittoq* a terrifying figure of "loneliness personified," a rejected figure doomed to wander the inland mountains and never dwell on the life-giving coast again. The names of *qivittut* were usually not returned to new children born in the community, a factor that makes their isolation, in the words of anthropologist Janne Flora, "more of a suicide, than suicide itself."[59] *Qivittut* left their homes behind, but only when home had become intolerable and unlivable. They often sought revenge on those who had wronged them. They were reported to move far to the north in old age, to a landscape where it was always dark—perhaps to a place like northern Ellesmere Island in December.

In the first few days after fleeing, *qivittut* could sometimes be rescued and reintegrated into human society. Later many acquired the ability to speak with animals, to run like caribou, to fly like birds, to turn invisible, or to see magically what was happening back home in their coastal settlement. Once they had supernatural talents, there could be no return to their family—at least not in the Greenlandic Christian tradition, which equated this with making a pact with the Devil and losing the Christian element of their soul. *Qivittut* were very rarely if ever "re-socialized."[60] However, there seems to have been faint hope. In one story, inland creatures helped a long-term runaway to return home safely: when she became tired, two dwarves "pulled the landscape together, so the distance became less to the land that she was going to, which was so far away it had a bluish color. The whole trip took only one day."[61]

Some West Greenlanders tried going *qivittoq* on nineteenth-century expeditions. Hans Hendrik, the veteran of four Arctic expeditions who lived in Angutisiak's home community, ran away from an earlier expedition in Lady Franklin Bay. His motives were multiple: he heard the men gossiping meanly about him; he thought he heard crew members discussing who should flog him; he missed his wife and children and friends; he was unaccustomed to military discipline. He chose to "go away to the wilds," deciding, "if I should freeze to death it would be preferable to hearing this vile talk about me." It is notable that he wrote *if* he should freeze to death. To the other expedition members, the question would presumably have been not if, but when. Hendrik traveled five miles and then stopped. He dug a hole in the snow near the ship and lay down in it, waiting to see if the others would come and look for him. They found him, and he returned to duty.[62]

When Angutisiak ran away in December 1881, was he trying to go *qivittoq*? It seems likely. Like other *qivittut*, he left without mittens or supplies—perhaps so others would assume he was dead, and would not suspect he was out there in the wind and snow, alive and seeking revenge. Commander Greely, who was familiar with Arctic literature, understood Angutisiak's flight as going *qivittoq*, although he seems to have been unaware of the revenge facet of these stories. Implausibly, Greely imagined that it was Angutisiak's childhood dream to become a "kivigtok."[63] Greely—who was then subsisting in an overcrowded cabin among men he disliked, and who came from a more individualistic society bursting with Romantic ideals of mountainous nature—envisaged a solitary existence inland as potentially preferable to life in an Inuit coastal community. But for Inuit, *qivittut* stories are horror stories about losing one's home forever. They are usually full of suffering and violence and grieving for everyone involved, including the runaway.[64] They serve in part as cautionary tales, warnings to avoid mistreatment of others. Nevertheless, the frequency of *qivittut* legends implies that there were individuals for whom going *qivittoq* felt like the best possible alternative. I suspect Angutisiak was one of these people.

While everyone on the expedition was dealing with an unfamiliar environment far from home, Angutisiak and Christiansen were also dealing with an unfamiliar culture. The absence of the sun may have disquieted Angutisiak, but it did not make him run away. Just as Gardiner's malaise was enhanced by the crowded winter cabin, so Angutisiak's unhappiness can be traced in large part to human factors. Greely did not believe that Angutisiak could feel mistreated. He proudly and repeatedly stated that his men all treated Angutisiak and Christiansen "in the kindest and most considerate manner." At the beginning of the expedition, he had ordered the men not to joke about or insult the new employees.[65] Yet racial stereotypes continued to be a basis for humor among the Americans. At the Christmas pageant in 1881, nearly every item on the program was an impersonation of some marginalized group. The main exception was Sergeant Jewell, who advertised his act as a "Select Reading," and then walked on stage and read the aneroid barometer to the appreciative crowd. Otherwise, the troupe put on an "Indian War Dance" and sang "Plantation Melodies." Then Private Schneider offered an unscheduled performance of an "Eskimo Belle," complete with fur clothing and a puppy instead of a baby on his back. It made the Qallunaat employees laugh raucously. According to Greely, Frederik Christiansen said the performance was "very good, very good," and Angutisiak cried. Greely assumed this was because it reminded Angutisiak of his wife.[66] Regardless of whether Christiansen and Angutisiak appreciated this performance, it seems inconceivable that stereotypes about "Eskimos" were not freely tossed around the expedition cabin.

Greely clearly viewed the two men with paternalism and some contempt.

He referred to Angutisiak in his expedition account as a "simple-hearted native," and to Christiansen, who had both Kalaallit and Danish ancestors, as a "wily and cunning" "half-breed." He assigned them segregated bunks, off to one side of the cabin.[67] Greely often wrote to his wife of his own homesickness, but he did not expect Angutisiak to be so tied to home and family, and was surprised to see "a child of the ice thus pine away."[68] He called the enlisted men by their last names, but he always referred to Angutisiak as "Eskimo Jens" and Christiansen as "Eskimo Frederick." It also appears that the soldiers could "volunteer" Angutisiak or Christiansen to accompany them on sledge trips.[69] When Sergeant Rice and Angutisiak went on a harrowing trip together, Greely named the strait they had visited "Rice Strait," a name that remains on Canadian maps.[70] Another man commented during that trip that he was "a little weary about Rice not returning," as if Angutisiak did not count.[71] Such racialized discourse and omissions were common at the time, but what Greely considered impeccably fair treatment of the Kalaallit was unlikely to have appeared so to them.[72]

Greely also wanted to believe that Americans were hardier than Inuit. When Angutisiak was exhausted on sledge trips, Greely considered it evidence that he had less "moral force and mental determination" than his American sledging partners. Yet since either Angutisiak or Christiansen was needed on nearly every venture far from the station, they were pushed more consistently than the other men.[73] Greely also commented that Christiansen lagged behind on the sledge trip on which Lieutenant Lockwood and Sergeant Brainard claimed the "Furthest North" record among polar explorers. He believed that Christiansen was "unable to appreciate" the purpose of this trip, which may well have been true.[74] Yet Brainard's field notes also indicate that the team had only two pairs of snowshoes. The unfortunate Christiansen did not get a pair, so he frequently broke "through the crust to his hips" and had to be dragged out by the dogs and officers.[75]

In December 1881, the month Angutisiak disappeared, Frederik Christiansen told Greely he believed the men were plotting to kill him.[76] True or not, Christiansen felt threatened. Inuit societies placed a very high value on getting along with others and not showing hostility, so a mild display of anger by American standards could be read as a warning of physical aggression and loss of control. In the expedition cabin, men raised their voices, showed anger, and criticized others in threatening language, both openly and behind their backs. Sergeant Cross—one of the more outspoken members who sometimes mocked Greely within earshot—commented in his diary after Angutisiak ran away that, "it was the opinion of most all of us that he ought to have a good flogging."[77] Surrounded by this kind of hostility, Angutisiak was reluctant to do anything to make trouble. One night on a sledge trip, when his sleeping-bag partner fell asleep before Angutisiak finished his work, he slept outside rather than disturb the American and got a frostbitten toe. Greely believed this was evidence of Eskimo Jens's "considerateness and kind heart," but Angutisiak may also have been using submissiveness to avoid anger and aggression.[78]

Before Angutisiak ran away, it had been obvious to many of the men that he was unhappy. Private Schneider wrote that Angutisiak was "downhearted" and "[showing] signs of fear."[79] Greely recorded in his journal that Angutisiak had repeatedly asked two of the enlisted men to kill him. Greely had tried to give him small gifts, but Angutisiak kept saying he did not deserve them, because he was bad.[80] I can't be sure why Angutisiak behaved this way. Perhaps these were self-deprecating gestures to people he viewed as aggressive and volatile, or perhaps he was also trying to be mistreated so badly that he would have clear justification for going *qivittoq*.

The morning that Angutisiak disappeared, Greely ordered four enlisted men to light turpentine torches, harness the dogsled, and launch a search. The men soon found Angutisiak's footprints, and they caught up with him in the early afternoon. He had traveled about twenty miles in a winding route, first south along a well-trodden path, then east and north. When the men found him, he was plodding slowly northward through falling snow, away from the expedition cabin and his home in Greenland. The men called out to him over and over again, but he didn't turn around; he didn't answer. When the men reached him, he surrendered passively and allowed himself to be put on the dogsled. He was given mittens, and fed some hard bread. For most of the way back to the cabin, Angutisiak said nothing. The only time he spoke was to insist a searcher run beside the sled rather than ride on it, for fear the man would freeze to death. In this half-day trip away from the cabin, one American seriously injured his shoulder, and another was brought back to the station in a frostbitten and delirious state.[81] It was a grim foreshadowing of the expedition's fate.

The Americans assumed that Angutisiak's actions were rash and senseless, that he was "a man who had deliberately turned his back on light, warmth, plenty, and comfort, to risk darkness, cold, want, and death." One private wrote that Angutisiak had "[taken] it in his head to go home or die."[82] But I do not think these were the only two options. Angutisiak was risking his life and humanity by walking away that day, but that does not mean he suffered from delusions about his ability to survive ill equipped and alone. The landscape he saw would have contained different possibilities and connotations than it did for the Americans.

Interlude: The Retreat

Before jumping ahead to the reappearance of the sun in 1884, when the men were starving to death two hundred miles to the south of Fort Conger, I will briefly explain how they ended up in such dire circumstances. A year and a half after Angutisiak's flight, in the continual daylight of August 9, 1883, the expedition members turned their own backs on the expedition cabin. They abandoned Lady Franklin Bay in their small steam launch, the *Lady Greely*.[83] They had spent 721 days at the station, 268 of which "had been marked by the total absence of the sun."[84] Resupply ships had failed to reach the men, who had had no contact with anyone for nearly two years. In retreating south, Greely hoped to find large stores of provisions left by the ships that had turned back.[85] Many of the soldiers, with considerable justification, considered it safer to remain at the station. It may not have been home, but it had become a relatively familiar place. They had little knowledge of the resources to the south, limited ability to transport supplies, and no information about the movements of the rescue ships. Everyone was still in decent physical condition. Greely had written earlier that winter that "nearly all the men weigh more eat more and look better than in 1881."[86] Even though the second year's diet had lacked some

FIGURE 3.6. The *Lady Greely*. The party used this steam launch to head south in 1883. LC-USZ62-136205, Library of Congress.

of the imported luxuries of the first, they had still eaten hearty dinners that consisted of soup, meat, vegetables, and dessert every night.[87]

The men left most of their gear and supplies at the station. In addition to the *Lady Greely*, they brought a dinghy and three other small boats. They carried only sixty days' provisions, four dogskin sleeping bags, and several three-man buffalo-robe sleeping bags, which expedition members had dismissed in the first year as "very cold and cheerless" in low temperatures.[88] They left their dogs behind; Greely did not want to kill them in case the party was forced to return.[89] The men were allowed eight pounds of clothing and personal gear; some chose to include treasured objects from home like family photographs. The officers took sixteen pounds each, and Greely attracted criticism for taking more than the allotted amount, perhaps as much as forty-eight pounds.[90]

Sergeant Cross, an alcoholic firebrand who despised his commanding officer, became increasingly venomous in his diary during the retreat. He stated that Greely did not know how to navigate, that he would not take advice, that he had refused to let men take extra warm clothes and skins from the cabin on departure, that he yelled at the men for no justifiable reason, that he burned blubber in the boiler that might be needed for food later, and that he spent his days huddled next to the heat source barking orders while the men were inadequately clothed and "some nearly barefooted." Cross's complaints culminated with the statement, "He has once or twice tried to regulate our bowels, but I think I have him there."[91]

Cross's grumblings were likely exaggerated, yet more sober voices corroborate his major grievances. Gardiner complained that the party had left cases of sheepskins at the cabin, which Greely had forbidden the men to sew into better sleeping bags. This was characteristic of expedition policy. The enlisted men, suffering with inadequate army-issue clothing and especially footgear, had been forbidden at Fort Conger from taking extra hospital blankets to sew into warmer clothes for themselves. Nearly all of them ended up purchasing blankets to make woolen underwear. When Gardiner had asked Greely if he could have a small piece of buffalo skin to sew a warm collar on his coat, Greely refused him even though, according to Gardiner, "we have many piece however which will never [be] used for any other purpose not being large enough." On the voyage south, Gardiner wrote, "Every one shaking with cold." He worried that Greely was rashly distributing rations with no concern for the future.[92] Sergeant Ralston concurred about the poor state of their equipment, stating that he and two sleeping bag partners had sewn pieces of sail to their bag to cover the holes, and that men were making desperate attempts to repair their boots.[93] At one point, unknown to Greely, the officers seriously contemplated a mutiny and a return to the station.

During the retreat, the expedition members were haunted and troubled by the unfamiliar sights and sounds of the ice. Sergeant Brainard was convinced that he heard the barking of Inuit dog teams, but after walruses bellowed in front of him multiple times, he conceded that he had mistaken one sound for the other. During one whiteout, Brainard wrote of the fluidity he felt between landscape and mental state: "The weather is still wretched in the extreme; the atmosphere is so thick and hazy that the coast cannot even be sighted; consequently we have no definite idea of where we are, or whether or not we are yet drifting with the pack. We are certain of only one thing, and that is our terrible sufferings. Everything else is indefinite."[94] Lacking adequate food or shelter from the weather, unable to get comfortable, fighting with each other, adrift physically and mentally: the men had not felt at home in Lady Franklin Bay, but they were now entering a worse place.

Strong winds and the drifting ice pack resulted at times in the party floating north rather than south. Eventually they were forced to abandon their steam launch, setting out across the ice on foot for the nearby coast of Ellesmere Island. They reached shore on September 30. Nine days later they decided to move north to Cape Sabine on the small Bedford Pim Island (today Pim Island), because Angutisiak and Sergeant Rice had located caches totaling thirteen hundred meal rations in the vicinity. Cape Sabine was likely the same location where, two decades earlier, Qillarsuaq's party had set out across the ice for Greenland.

Here I want to briefly consider the retreat of the Lady Franklin Bay expedition in the context of Arctic travel. First, Inuit travel. Extreme voyages like the Qillarsuaq migration could experience great hardships, but there are far more examples of Inuit long-distance journeys that were neither desperate nor strange occurrences. In the 1850s, a whaling captain met an Inuk one year near Cumberland Sound; the following summer he was surprised to see the same man at Pond Inlet, over five hundred miles north.[95] In 2008, Elder Jamesie Mike described to me how Inuit used to routinely travel between these two regions. "By dog team," Jamesie explained. "By dog team, not by whaling ship." He wanted, I think, to make sure that I understood that long-distance travel was an Inuit tradition that predated whaling. Inuit could travel hundreds of miles just to visit, Jamesie related. "It would take a very long time. There was no hurry, and along the way you'd have to hunt and feed the dogs, so it would be slow." He continued, "They would have taken somebody who had done the trip. There would be absolutely no maps; it would all be from somebody's memory. The only time any family group or anybody would go on these long-distance travels is if there is someone in the group who has taken the trip before, so they would have a guide." Jamesie's description stands in stark contrast to the retreat of the Lady Franklin Bay expedition: in Jamesie's account, entire families traveled together; they lived off the land; they went only along known trails. The journey may have been new and exciting to most of the travelers, but the paths and connections were part of collective history and experience.[96] While very few Qallunaat ever reached this level of comfort with Arctic travel, many followed similar protocols. Qallunaat whaling captains often returned to the same harbors; expedition leaders tried to hire officers with Arctic experience; and countless Qallunaat from police officers to scientists to ministers have relied on local Inuit guides and hunters, who have their own versions of these events. The Lady Franklin Bay expedition was a spectacular failure in terms of local knowledge.

At Cape Sabine, the men built a low dwelling of stone walls three feet high, roofed with an overturned whaleboat and canvas. Some of the taller men nearly touched the roof when they even sat up. With twenty-five people and their essential gear inside, the party sat elbow-to-elbow with only a narrow passageway and cooking space in the center.[97] The dwelling was cold. Sleeping bags froze solid whenever the men vacated them, and ice glued the bags to the ground for seven months. The medical report from the rescue party's doctor said the temperature of the hut hovered between 5 and 10°F (−15 to −12°C).[98] The crew complained more about hunger than cold in their diaries, but they must have been freezing. Starving people often feel unable to get warm, even in the summer or under layers of blankets. Other effects of famine include irritability, emotional instability, and depression.[99] In a statement that captures the dejected mood, freezing temperatures, and general discomfort that suffused the hut, Greely ordered that all men were to keep their heads out of their sleeping bags between breakfast and dinner.[100]

Over the winter, the party began to starve and deeply suffer. An Arctic army ration was forty-six ounces of solid food per man per day. On November 1, after the sun had set for the year, the men redivided their rations to a daily issue of less than 15 ounces.[101] That same month, Corporal Elison was carried in from a trip to a nearby cache, his "feet and legs frozen nearly to the knees." Despite concerted attempts to treat him, he lost the use of all his fingers and thumbs. His foot fell off in January, although the tissue had been dead so long that he failed to notice. Elison later asked to be killed so that his rations might be saved for the others, but they refused, and even continued to give him extra bits of food and chocolate from their very limited stores.[102] On the solstice, December 21, Sergeant Cross wrote, "Thank God this is our longest darkest day. We will have the sun commencing to come back tomorrow."[103] Cross was the first to die a few weeks later, of undernourishment and perhaps scurvy. The others all lived long enough to see the sun's return in February, but their resources were severely depleted. They probably had no idea how long they could go on living. Many American scientists of the era purported that humans could survive no more than ten days without food.[104] And after the sun's return, some members of the party would be called on to assume a disproportionate level of risk to save the others.

The Reappearance of the Sun in 1884

The sun returned to Cape Sabine on February 17, but it was not visible from the hut. Greely commented in his journal that just knowing the sun had returned cheered up several of the men. Kislingbury called it "the good old friend."[105] People showed more optimism and hope.[106] Yet they were disturbingly frail. George Rice, an energetic grey-eyed man in his late twenties, was perhaps the strongest member of the party at the time. He noted in early March that no one had yet walked to a point where they could view the sun. They were too weak, and their rations too paltry "to allow [for] any exertion not absolutely necessary."[107] Three days later, Rice climbed to the summit of a nearby hill and viewed the sun.[108]

On March 11, the sun shone on the hut for the first time.[109] When David Brainard saw his first sunrise, he wrote in his diary that he "lingered among the rocks to enjoy the warmth and comfort which it brought. To use the apt expression of [Arctic explorer] Dr. Kane, 'It was like bathing in perfumed water.'"[110] In late March, the men made some holes in the boat roof and covered them with canvas, so they could see each other in natural light again.[111] On April 24, they stripped more of the boat and took the snow off part of the roof. Sunlight flooded through the cracks and holes, and rays reflected off some tin cups inside the hut. The frostbitten Joseph Elison "thoroughly enjoyed the scene and brightened up considerable both in spirit and temper." Another man was grateful that the sun was shining brightly and "the backbone of winter [seemed] to be broken for good." By May, the men were placing a sleeping bag outside during the day so "those able to crawl out" could engage in "a glorious sun bath."[112] At times the sun provided them with a different kind of bath, as it thawed ice that had accumulated in the hut. Clothing became saturated. "The melting of the frost from the roof renders our condition positively wretched," wrote Brainard.[113] Greely commented that his first thought on seeing the "utter squalor and misery" of the hut in natural light was, "How have we ever passed through this hell on earth and kept our reason?"[114] Nevertheless, most if not all the men felt their spirits lift from the return of the sun. It brought back hope.

That spring the party became viscerally aware of the most fundamental human connection to the sun: its ability to produce and sustain life. At Fort Conger, the Qallunaat had welcomed the lengthening days mostly as a chance to bask in natural light, to escape the cabin, to get out on exploratory sledge trips, to collect and press local plants, and to eagerly await the arrival of a ship.[115] Many had tried their hand at hunting and fishing, but their interest in wildlife had been largely scientific and recreational. Now, everyone became desperately focused on sustenance and on the resources the sun might provide. The stockpiled rations had almost been exhausted, making the party reliant on what they could hunt and gather. They were living off the land.

Just as the strengthening sun grew crops in the United States, so it heralded an intensifying of life at Cape Sabine. An Inuktitut definition of the season of spring or *upingaaq* reads in part, "the sun warms and all the animals, the seals, the birds, the fish, all begin to move, and humans too."[116] Hunting became easier. "Before Sun returned only five hundred pounds meat obtained," wrote Greely in his first major correspondence with Washington after the rescue.[117] In mid-March the party saw black guillemots, and seals became more numerous. The men also began to net small crustaceans. "All are much delighted at the game and feel much encouraged," commented Greely.[118] Other birds returned; a snow bunting sang on the roof of the hut in early April.[119] These migratory birds and the odd raven were eagerly devoured whenever they could be shot. The birdsongs also connected the men to home, "broke the silence and gloom," and reminded them they were not alone in the world.[120]

The greatest triumph occurred on April 11. As several members of the party lay on the verge of death, Angutisiak and Private Long pursued and killed a polar bear.[121] After ensuring the bear was dead, they knelt and gulped its "inspiring and life giving blood." When they returned to camp with the good news, Angutisiak rushed to the bedside of Elison and "with a voice full of emotion told him, 'You all right now Elison.'" The few men who were still strong enough went to fetch the bear; they "carefully chopped from the ice every bit of the precious blood."[122] The butchers passed small pieces of meat and fat to the other men. Sergeant Brainard wrote, "What words are adequate to express the rejoicing manifest in our little party tonight? . . . ere many months we will have returned in safety to our homes. . . . Life is now ten times sweeter than at any former period of our existence."[123] He and others felt that the bear signified their salvation, that it would get them home. This proved to be overly optimistic, but the bear did provide three hundred pounds of meat, and nothing was wasted. Private Long shot a seal the following day.[124]

As in most Arctic regions, it was sea mammals that offered the best chance of survival. Seals were far more plentiful than bears. Back in Lady Franklin Bay, Angutisiak had harvested whatever was available, but had taken particular pride in his sealing, which would have been his primary hunting activity back home.¹²⁵ Many men had resisted eating seals at Fort Conger, but they began to applaud the hunt on the journey south. One expedition member noted on the retreat that "those who disliked [seal meat] before ate it with relish." Another wrote that "even the blood is now considered a luxury, and is eagerly sought for by almost all of us." The "almost" probably referred to Sergeant Ralston, who commented to his diary that the "vampires were turning out for a drink of blood."¹²⁶ For most of the party, however, eating seal meat and blood was an easy and life-saving adaptation.

The Kalaallit employees were both skilled sealers, and they had brought a kayak on the retreat. On their own, they would have stood a far better chance of survival than any of the Americans. Yet their hunting skills did not save them. If anything, they put Angutisiak and Christiansen more at risk. They were expected to hunt for the party, and hunting was an exhausting and dangerous endeavor.

Christiansen was the second member of the party to perish. On March 18, the doctor recommended that Christiansen not be sent out hunting anymore; he was too exhausted and his feet were swollen. A few days later Angutisiak's feet also swelled up. The doctor recommended letting both Kalaallit rest, but Greely insisted they hunt on alternate days. On March 25, Christiansen was "half-carried" in by his hunting partner, Private Long. Six days later he fainted, and despite being given extra food that week, he became delirious and died on April 5.¹²⁷

Shortly thereafter, Greely confided to the hospital steward that he had begun issuing himself an extra two ounces of rations per day "as he saw the necessity of keeping himself up for the well of the party." Greely was forthright enough to mention this in his official report, arguing that he did it selflessly, out of a not-unfounded fear that the party might fall into worse hands if he died. Still, those extra rations contributed to Greely's eventual survival. Life and death were as much about hierarchical relationships as individual fortitude.¹²⁸

Of course, individual strengths also helped determine who lived and died. If Angutisiak and Christiansen had little choice but to provide for the party, some Americans chose to do so. The unfaltering Private Long performed far more than his share of physical labor and survived. Sergeant George Rice, a congenial and charismatic photographer who had sailed north in 1881 with love letters from many different women, consistently volunteered for the most exhausting and dangerous duties. He died only four days after Christiansen. Rice collapsed from exhaustion while trying to retrieve food from a cache. He had worked himself to death. His traveling partner reported that on Rice's last day, "his mind continually reverted to home, relatives, and friends, and to the pleasures of the table on which he intended to indulge upon his return."[129]

Rice and Christiansen's deaths made the party even more dependent on Angutisiak, and daily hunting took its toll. As the weeks passed, Angutisiak became exhausted and pitiful. Described just before his death as a "faithful and indefatigable worker," he had begun to lament aloud that "Eskimo no good."[130] He was weak, constipated, and "in bad spirits"; he complained to the doctor of feebleness in his legs. He and Private Long were issued extra pemmican to keep up their strength as hunters.[131] Angutisiak spoke repeatedly about his wife and four children, and how much he longed to see them.[132] The other men, reliant on his hunting, began to promise him all kinds of things on their safe return home: a boat, a watch, a new kayak. They were anticipating a time when their wages could once again buy food. Shortly before his death, Sergeant Rice sat gazing at a bank bill. His comrade commented, "What a bitter mockery in our present situation, to look at money, when we have nothing to eat."[133] The Qallunaat equation of money with a good home no longer added up.

Francis Long and Angutisiak continued to hunt together until April 29, when Long returned with tragic news. They had spotted a seal sunbathing on a pan of ice, and Angutisiak approached it in his kayak. Sharp ice cut the skin boat. Angutisiak drowned "without uttering a single cry for assistance." Long tried and failed to reach him over the surrounding ice floes. Angutisiak's body sank from view. This event may have held a particular horror for Long, since several weeks earlier Angutisiak had risked his own life to rescue Long from an ice floe, reportedly calling out to him, "You go, me go too!"[134] Angutisiak's death, and Long's inability to help him, reflects the disproportionate level of risk that Inuit employees often assumed on expeditions. They faced perilous situations while others, sometimes desperately, looked on.

The news of Angutisiak's death was a "terrible blow" to the men.[135] It is unclear how many of them could have been called his friends. Yet they all mourned the loss deeply. They knew that Angutisiak's skills, especially in a kayak, were keeping them alive. One man tried to make light of the news, saying it was "bad news" but that "no one is discouraged yet."[136] Others disagreed. Gardiner described the death as a "heavy loss." Sergeant Brainard feared it would "prove fatal" to them all.[137] "The death of Jens has left us in a bad predicament," summed up one private, "as we have no chance whatever to procure a seal in the water [without a kayak]." In desperation, Long—who was an excellent shot—continued to kill seals, hoping their bodies would drift toward the shore or a reachable piece of ice.[138] They never did. The seals that could have saved lives were out of reach.

Angutisiak was survived by his wife and four children, and Greely requested that his widow receive a pension of four American dollars per month, half of what the widows of American privates received.[139] Drowned at sea, his body was never returned to his community, although news of his death was. In 1886, his brother's family baptized their new baby Jens Edvard. It seems that the name, or *atiq*—which also contained parts of the dead man's identity—had been passed on.[140] In this way, Jens Edvard Angutisiak returned home to Kangersuatsiaq.

Summer Solstice 1884

On June 21, 1884, David Brainard made the last entry in his expedition diary. He noted that it was the longest day of the year, and that Private Connell could no longer use his legs below the knees.[141] By now there were only seven men left. Sergeant Hampden Gardiner was not among them, having died days before, clinging to life for two months after the doctor predicted his demise. Greely wrote that Gardiner had lived on through sheer force of will, apparently motivated by his "intense desire to return home" and see his family again. On his last morning he had struggled partly out of his sleeping bag, clutching family portraits in his hand. His last words were, "Mother! Wife!"[142]

The remaining men were all on the edge of death. Some of them now fainted "from sheer exhaustion" after moving their bowels, and were so weak they had to defecate indoors.[143] They had dismantled their dripping hut and burned the boat-roof for fuel. They were now living in a canvas wall tent, which was partially collapsed and full of holes that the wind blew through.[144]

The blustery morning after the solstice, the dying men heard a sound like ship's steam whistle, but saw no ship. They discussed the noise and decided that it was probably just the wind blowing on a tin can. Francis Long, ever hopeful, stumbled up the hill and saw a black speck in the distance. He raised the "flag"—some old underclothes tied to an oar—and saw a steam launch approaching the shore. Then he "tumbled rather than ran" down the hill to meet his saviors.[145]

As the rescuing party approached the tent, they saw what they thought was a corpse, which turned out to be Maurice Connell, "cold to the waist" and barely breathing. Inside were five men in rags, covered in filth, skin hanging "in flaps" from their bodies. They emitted a "sickly, offensive odor" of stale urine and had "wild and staring" eyes. One of them, Elison, had missing fingers and blackened remains of feet, with a spoon lashed to the stump of his hand. The men were loaded onto the ship. Four of them were unable to walk, and Greely later stated he felt they were within a day of death. Elison died on board. The other six men returned safely to the United States.[146]

FIGURE 3.7. The tent from which the survivors were rescued. National Archives (US) photo no. 200S-LFB-71.

The rescue of the surviving Lady Franklin Bay expedition members—and the deaths of the others—created a sensation. On July 18, 1884, all major American newspapers led with the rescue, and extensive coverage continued for weeks. The story appealed to a wide audience already familiar with Arctic disaster narratives. As historian Beau Riffenburgh summarized, the Lady Franklin Bay story had everything Qallunaat readers craved: "hardship, scurvy, a farthest north, a desperate retreat by boat and sledge, three winter ordeals, suicide, insanity, and death."[147] The survivors' riveting accounts shaped how Americans viewed the Arctic, but they also overshadowed contradictory narratives from other expedition members and from elsewhere in the Arctic world.

The last months at Cape Sabine had been grim. By May 17, the men had consumed all their rations and were subsisting solely on what they could gather from the land. It wasn't much. With no kayak or other watercraft, the only marine resources they could access were in the tidal zone. Ingeniously, they found a way to harvest small crustaceans that they referred to as "shrimps" or sometimes, less optimistically, as "sea flies." These creatures, *Onisimus edwardsi*, were described by a crew member as being "about the size of half a grain of corn."[148] One of the men counted that thirteen hundred shrimp fit in a half-gill measure, about a quarter-cup.[149] "Shrimping" was a painstaking task, largely borne by Sergeant Brainard from the time he first proposed it in mid-March.[150] He collected these crustaceans by lowering bait—bearskin, seal bones, sealskin clothing—through the ice, pulling them up repeatedly, and shaking off all the animals.[151] Each person's daily portion was meticulously weighed out, and often it "would barely cover the hollow of his hand."[152] The men had to force the shrimp down. In the words of Private Long, "they were very gaggy."[153] Still, the crustaceans provided a reliable source of sustenance. They supplied nearly half of the calories consumed by the expedition members for the last three months of their stay.[154]

But the "shrimps" were not enough to sustain life. The men began to eat their sealskin clothing rather than use it as bait.[155] At the end of May, one private, against Greely's orders, roasted and ate his sealskin boots.[156] Others sampled caterpillars and ptarmigan droppings. They picked and ate saxifrage greens. Nothing that could possibly provide nutrients was passed up. Sergeant Brainard wrote, "Crumbs of bread which are occasionally exposed . . . through the melting snow are picked from heaps of the vilest filth and are eaten with avidity and without repugnance." After some debate over whether lichens were poisonous, the party began to boil black lichens called rock tripe or *Umbilicaria*. When stewed, the rock tripe "made the water resemble thin mucilage."[157]

The men spent vast amounts of time at Cape Sabine talking about the food they were not eating. Any meals that Angutisiak and Christiansen reminisced about were not recorded, but the Qallunaat spared no words. They spent hours concocting delectable menus on which they hoped to feast after they were rescued. Although they had eaten some lavish dinners in their first years in the Arctic, their hunger led them back to the United States. They planned to consume these meals with old friends, or to invite their current sleeping-bag mates to dine with them at home. Sergeant George Rice recorded many of these banquets in his diary. After a disappointing breakfast in March, he wrote simply, "I must now record my imaginary bill of fare. Place, Washington. Breakfast, broiled mackerel with baked potatoes, mutton chop and poached eggs on toast, apple pie and chocolate."[158] After the rescue, Maurice Connell claimed that he "used to lie awake thinking of food until [he] was crazy," and then he would fall asleep and dream about it. In one dream, he snatched a fresh-baked loaf of bread from a kitchen shelf, then woke up and found himself "grabbing the empty air" inside the hut.[159] Waking life was a nightmare from which the men could not escape; their reality was the antithesis of home.

Memories of home, and especially of family, were not just welcome distractions in these final days; they could keep starving people alive. Gardiner appeared to survive off longing alone for his last two months, and Brainard, the shrimper, wrote in his diary that thoughts of home and future sustained him. He was also strongly motivated by the other starving men, "the poor fellows, who . . . look to me to provide them with food." Whenever he thought of giving up, the "faces of [his] friends and family" appeared before him, as if "reproaching his weakness," and he persevered.[160]

The rescue ship was a long time coming. Men died "by inches" during the last three months of the expedition, with a brief respite in late April and early May due to the influx of game.[161] Death became an everyday matter rather than a terror. After witnessing many deaths by starvation, several men concluded that it was "not so painful or terrible," especially compared to their own daily sufferings. Just before death, the men generally lost all appetite and began calling insatiably for water; later they became unconscious.[162] Often the process of sliding into death was so gradual that it was unclear exactly when it happened.[163] Sergeant Brainard recorded that everyone became "utterly indifferent . . . to the presence of death."[164] When his friend Lieutenant Lockwood was among the first to die, it "affected him deeply" to pass his grave and see the buttons on the officer's shirt sticking above ground and glinting in the sunlight. "But this feeling soon wore away," Brainard reportedly said. "We had so many other horrible things to think of."[165] A few weeks later, men would die in shared sleeping bags and their comrades would only muster the energy to move them the following day.[166] Later, on board the rescue ship, Brainard stood quietly, watching the galley scrapings being emptied down a chute. When asked why he was staring, he replied, "I have seen enough good food thrown away since I have stood here to have saved the lives of our nineteen dead."[167] Through heartbreaking comments like these, the United States became linked even more to abundance, and the Arctic to desolation and suffering.

In the end, the men simply could not eat enough lichens, crustaceans, and moldy crumbs to stay alive. The rescue party found that some of the bodies had been neatly cut into and butchered. Recent nutritional research has indicated that all the men would have died without an additional source of food, so at least some of the crew almost certainly ate the dead.[168] The survivors forever denied their complicity. Nevertheless, the *New York Times* broke the story with a front-page headline: "Horrors of Cape Sabine: Terrible Story of Greely's Dreary Camp. Brave men, crazed by starvation and bitter cold, feeding on the dead bodies of their comrades."[169] In the days that followed, other American papers offered "Further Facts about the Ghastly Prison in the Arctic Seas," and "Something New and Sensational in the Line of Arctic Horrors." Bodies from the expedition were exhumed in the United States and indeed proved to be less than whole.[170]

The cannibalism story opened a national debate, but many if not most nineteenth-century Americans sympathized with the survivors. Readers were appalled that New York reporters victimized men who were only just recovering from starvation. They stated that in a similar situation, they would want their own bodies eaten to prolong the life of their companions. Supporters also justified the cannibalism because of the men's surroundings. One doctor wrote, "It is all very nice for us down here, able to procure all the necessities of life . . . to shudder at these reports." Another writer commented, "You can't judge men in their condition as you could if they were in civilization. Just think of it—going for months without having water to wash their faces in."[171] Cannibalism was portrayed, here and in other southern accounts, as a potential consequence of venturing into such a barren, unhygienic, and uncivilized environment.

Inuit have also spoken of cannibalism with a mixture of horror and pragmatism. As the Inuk interpreter Andrew Dialla put it to me, "Cannibalism is wrong, people don't like it, but we know it's necessary sometimes."[172] Instead of declaring the land inadequate, Inuit stories tend to focus on people's limited mobility: what factors prevented starving people from catching game, or from moving on to a richer place? Why could they not contact other groups to help them?[173] Elder Saullu Nakashuk explained in one story, "When they starved it was because the animals they were expecting to be there or go there did not go there."[174] Inuit homelands can provide what is needed, but only if people are free to move around in them, and only when they can correctly anticipate the movements of animals. Americans critiqued their government's failed attempts to resupply and rescue the Lady Franklin Bay party, but this is about movement in and out of the Arctic region. Far less attention is given to the local—but no less important—loss of mobility when the party lost Angutisiak and his kayak, and burned their boats for fuel.

What of the survivors themselves? They returned home, and physically they recovered quite well. They spoke proudly of their accomplishments: the Furthest North record, their scientific observations collected even as they starved, their proven ability to survive and provide for others.[175] If they were haunted by personal failures and doubts and flashbacks, they did not say so. The federal government would in future shy away from funding large-scale polar expeditions, but the press and private funders did not. The enlisted men were promoted, and most went on to have successful careers in the army or other government employment.[176] David Brainard, the shrimper, ended up a Brigadier General. Adolphus Greely retired as a Major General and received a Congressional Medal of Honor for his lifetime of military service. Both he and Brainard are buried in Arlington Cemetery.[177] Francis Long, a German, became a naturalized citizen of the United States in 1888. He married and had three children. He named one of them Sabine, after the cape where he had suffered and been rescued. Alone among the survivors, Long embarked on another polar expedition.[178]

*

There had been fourteen expeditions for the first International Polar Year, including a German station in Cumberland Sound. Only Lady Franklin Bay— the sole venture to end in disaster—was widely reported on by the Anglo-American press. The story was brought home. Newspapers and periodicals, which had generally become more sensational since midcentury, continued to cover the survivors' ordeals, their recovery, their public appearances. Greely made several lecture tours and wrote popular articles. Sergeant Rice's expedition photographs toured in lantern slide shows. A decade after the rescue, a "cyclorama" from the Lady Franklin Bay expedition was displayed at the 1893 World's Columbian Exposition in Chicago. A huge painting and model icebergs were mounted inside a curved surface to immerse the viewer in the scene. Live human actors stood in front of the painting. Several different stereoviews of the cyclorama were produced for viewers at home.[179]

At least one survivor participated in a drama at the Union Square Theatre in New York. Entitled *Storm Beaten,* the show dealt "with the Arctic regions and its frozen horrors." Four of the men toured and gave short lectures for a thousand dollars a week.[180] One of the survivors, Maurice Connell, spoke in an interview of being "back in civilization," and the reporter commented: "Mr. Connell, in common with the other survivors, has learned to distinguish between populated countries and the isolated Arctic regions as if they were two different worlds, and invariably refers to the former [as 'civilization']."[181] That statement reflected popular American ideas about the Arctic as a counterpart to the welcoming temperate world: it was a terrifying, separate, empty place that could strip anyone down to skin and bones.

The Lady Franklin Bay expedition continues to captivate audiences. Its story has been retold as a heroic tale of self-sacrifice, a tragedy, a scientific quest, and a cautionary tale. It has been featured in articles, books, and documentaries.[182] The survivors' writings remain gripping, and there is no denying the fortitude, compassion, and inventiveness that some of the men showed in the face of misery. Yet many accounts of this and other polar expeditions reinforced existing stereotypes about the Arctic as an alien, "unhomelike" counterpoint to the United States. They overshadowed countless other tales being told in this period by Inuit and Americans, who were crisscrossing Arctic landscapes by ship and whaleboat and kayak and dog team. Through repeated retellings of gruesome events like Greely's final months at Cape Sabine, American conceptions of the Arctic became haunted by the specter of seven incapacitated men starving in their own filth, cutting flesh from the bodies of their dead friends. The United States was a "natural" home; the Arctic a place that could never be home. It was a place where even skilled hunters like Angutisiak could not survive, and where good men like Hampden Gardiner did their best but still lost their lives, where they cried out for home into a vast silence that was "as calm and still as death."

4

Inuit in Cumberland Sound

In 1994, the Inuk Elder Aksayuk Etooangat sat down with the young woman Margaret Nakashuk in the hamlet of Pangnirtung in Cumberland Sound to record some of his stories and legends.[1] Etooangat began with a story his father Angutiqjuaq had told him about his first encounter with Qallunaat, probably Scottish whalers. These events occurred near Qatiggait fjord, which Etooangat had later been able to find and visit because his father had shared the place name with him. In Qallunaat terms, the meeting took place in the mid-1800s, on Padloping Island off the Cumberland Peninsula, 125 miles north of Cumberland Sound. Etooangat explained: "When [my father] was a boy he never saw any white people. . . . He was only living the traditional Inuit way of life with absolutely no contact with Qallunaat. Finally a ship arrived. But nothing bad happened and nothing bad was done to them. And Inuit started going visiting to the ship, and the ship people welcomed them." The Qallunaat gave the Inuit a small barrel, a box of matches, and a clock. The barrel contained tobacco; it stank and tasted horrible. Etooangat's father kept the barrel and the matchbox, which were useful containers. He dumped their contents overboard before he even reached the shore. All that fall, tobacco washed up along the shoreline. Later in life, Etooangat recalled, his father would yearn for tobacco and think back to when he had thrown a whole barrel of it away. The Inuit eventually threw the clock into a nearby pond, because it disturbed them by making a sound like a heartbeat when it was not a living thing. Etooangat asked, on the tape, if Inuit who were traveling in the Paallavvik area could try to find the pond. He summed up his father's story by saying, "That was the very first time they saw Qallunaat, and the very first time they received gifts. They basically threw them all away."

FIGURE 4.1. Etooangat in Pangnirtung, 1990. NWT Archives/Northwest Territories. Department of Public Works and Services/G-1995-001: 6561. Photo by Tessa Macintosh.

While most Qallunaat narratives about the Arctic emphasize the foreignness of Inuit culture and northern landscapes, Angutiqjuaq's contact story highlights the foreignness of the Qallunaat and their possessions. Inuit were self-sufficient: they lived "the traditional Inuit way of life with absolutely no contact with Qallunaat." They made fire without matches; they lived well without tobacco; they had no need for clock time. They were at home in this place.[2]

Angutiqjuaq was one of many Inuit who followed the foreign whalers into Cumberland Sound. The availability of trade goods—often assumed to be the major draw of Qallunaat ships—does not appear to have been his main motivation. He packed up and left to work as a whaler after his in-laws criticized him. Angutiqjuaq must have lived through most of the commercial whaling period in Cumberland Sound, as he was too old to be the family's main hunter by the time Etooangat could remember him in the early 1900s. As Etooangat put it, his father "lived to an old age, and he lived through all the old ways, and then he worked as a whaler with the whalers."

Few Inuit stories like this one about the early years of commercial whaling have survived, but many Elders from Cumberland Sound have recorded their memories of its end in the early twentieth century. In sharing parts of these stories in this chapter, I follow different conventions than elsewhere in this book. Inuit are generally very careful to explain if they did not experience an event personally and to name the deceased person whose story it is. The

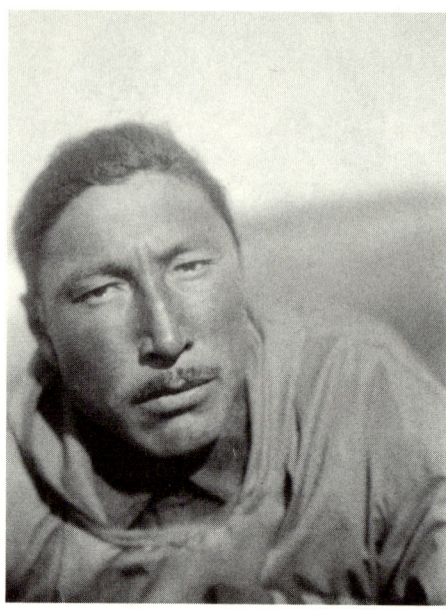

FIGURE 4.2. Etooangat in Pangnirtung, 1936. A. G. McKinnon/Canada. Dept. of Indian and Northern Affairs/Library and Archives Canada/PA-102076.

context in which a story was told, and to whom, can also be critical. Academic writing often relegates this type of information to footnotes, but I incorporate more of it in the text in this chapter, especially the distinction between first- and secondhand knowledge. Furthermore, the ways that Inuit Elders situate stories in time and space will be unfamiliar to many Qallunaat. Instead of, "in 1909," a storyteller might say, "when I was old enough to remember, but still young enough that my mother held me by the hand to take me places." Instead of, "in North Pangnirtung Fiord," an event might take place "at Ukialliviminiq," a specific site near the mouth of the fjord where, as the place name conveys, people used to live in the fall while they waited for freeze-up. Details can also place a story at a certain time of year. I include many Inuit markers of time and space in this chapter, partly so Inuit readers can accurately situate the stories—and so Qallunaat can too: many Inuktitut place names have recently been made official, or will be soon. I also retain markers so Qallunaat readers unfamiliar with Cumberland Sound can consider the extent of their own foreignness as well as how much there is to understand. Overall, however, I still retell stories in a Qallunaat style, using quotes and summaries in an academic narrative. The range of elegance and meanings of Inuit stories can only be grasped in original Inuktitut versions, especially those told repeatedly at length, in depth, and without interruptions.[3]

Up till now, this book has focused on people far from their homes. This chapter aims to provide a glimpse, albeit incomplete and translated, of what it meant to be at home in a specific part of the Arctic. Cumberland Sound during commercial whaling was a place where Inuit were working alongside Qallunaat, smoking tobacco, consulting shamans, eating hardtack, participating in global commercial economies, and continuing to hunt seal and caribou as they had always done. Qallunaat are part of Cumberland Sound history in this period, but, at least as importantly, there are countless Inuit stories where they do not play a role at all. From Angutiqjuaq's time to the present, Inuit have welcomed some Qallunaat individuals, goods, and traditions into their homes. Yet, like Angutiqjuaq, they have continued to refuse many supposed "gifts" from Qallunaat. Through commercial whaling and beyond, Inuit have pushed back against unwanted Qallunaat incursions, for the most part subtly and cautiously, to uphold their own ideas of home in Cumberland Sound.

Inuit born near the end of commercial whaling lived through devastating changes instigated by Qallunaat. But Elders' stories were not all about upheavals. For many Inuit, continuity and change are not opposed. As hunting people in a region where animals, ice, and weather are constantly moving and shifting, Inuit have long valued patience, flexibility, contingent but deep local knowledge, and learning to anticipate and adapt to the unexpected instead of trying to control it.[4] Elders also spoke about what Inuit have held onto: hunting and whaling, families and communities, and connections between Inuit and the land. While acknowledging drastic and often painful transformations, they showed both how the land remained home and how they have kept their homeland recognizable.

Like chapter 1, this chapter runs through six Inuit seasons. It begins where the Inuit year traditionally does, during the season of *ukiuq*.

Ukiuq

Qatsu Evic grew up on Qikiqtat, the island whaling station where Etooangat also lived with his parents. In her old age, she recalled an event that had taken place there in *ukiuq*, at the darkest time of the year. When Qatsu was about five years old, a Qallunaat cooper stole some alcohol from a nearby ship, drank it, and died. "I saw it," Qatsu stated, "the foam from the mouth reached to the floor."

This event was recorded by the local missionary, who returned from a dogsled trip in January 1905 to find the man dead in the whaling station house. The cooper's name was Davidson, and it was his first winter in the Arctic. Inuit reported discovering the body after noticing the lack of smoke coming from the chimney. The missionary, Edgar Greenshield, listed the cause of death as a stroke, with no mention of alcohol. His story ends here, but Qatsu's version continues on.

Soon after Davidson's death, Qatsu's father went out hunting. Back at home, dogs began to bark. Qatsu's mother Aasivak heard footsteps outside, and the dead cooper flung open the door. His face was white with frost. Aasivak yelled to her daughter to wake up. She charged at the door and the cooper backed away, but still she could hear his frozen skin *kamiik* boots walking around outside. Then suddenly the noise stopped. "This happened after he was dead," Qatsu concluded. "[My mother] told me that story the next day."[5]

This story is still known and retold in Cumberland Sound.[6] It is a good reflection of the terror that Inuit children of Qatsu's generation often felt even for living Qallunaat—and, I think, of the fear an Inuit woman could have felt hearing unwelcome Qallunaat footsteps approach when her husband was not home.[7] But it also speaks to another facet of life at the whaling stations: how Qallunaat could haunt and hang over Inuit homes, but could sometimes be shut out from them.

In Qatsu's youth in the early twentieth century, Cumberland Sound was no longer dotted with the American and Scottish ships described in chapter 1. Full overwintering crews were usually the unplanned result of a ship breaking down or becoming trapped in the ice. Whaling and trading companies now contracted out the whaling to Inuit, sending a ship each summer to pick up the proceeds. The bowhead whale population had plummeted in the intensive hunts of the mid-1800s, so Inuit also traded furs, blubber, and ivory from other Arctic species. Qallunaat employees—usually just a manager and perhaps a cooper—lived in wooden houses surrounded by Inuit communities at two main commercial whaling stations on opposite sides of Cumberland Sound, Qikiqtat (Kekerten) and Uummannarjuaq (Blacklead Island). After 1894 there was also an Anglican mission at Uummannarjuaq. Some years, however, Inuit managed the station activities and there were no Qallunaat overwintering in Cumberland Sound at all.[8]

Qallunaat activities, or lack thereof, still affected Inuit lives. The populations of the whaling stations fluctuated, but at times there were communities of over a hundred people at Qikiqtat and Uummannarjuaq. These were large groups by Inuit standards. At the stations, Qallunaat diseases weakened and took many lives, especially in the weeks following the arrival of a ship. Many Inuit did jobs for Qallunaat or Qallunaat enterprises for most of the year. In addition to hunting and processing animals to trade, they worked as seamstresses and guides, sometimes putting their own lives at risk to accompany

FIGURE 4.3. Qatsu's mother Aasivak decades after the haunting, hanging *kamiit* (boots) to dry in Pangnirtung in 1946. George Hunter/National Film Board of Canada/Library and Archives Canada/PA-166445.

Qallunaat determined to travel.[9] Some Inuit women also had relationships— either short- or long-term, consensual or not—with Qallunaat men.

Inuit were very poorly compensated for their labor at the stations, given that the baleen of a single bowhead whale could fetch $10,000 at the turn of the twentieth century. They generally received basic rations of supplies like hardtack, coffee, syrup, and tobacco each Saturday, with bonuses such as a gun or a boat at the end of the season, or when they caught a whale. In the 1970s, the Elder Markosie Pitseolak said, "Today we can only imagine all the money we held in our hands [when we butchered a whale.]" He added, "But because we were used to that way of life, it did not seem like we were being robbed."[10] Profits flowed south to owners and investors, but they were not apparent in Cumberland Sound, where Qallunaat residents lived fairly meager lives.

David Cardno, the lone Qallunaaq at Qikiqtat whaling station during the First World War, gratefully recalled his everyday interactions with Inuit. They "built him in" to his drafty house by constructing an insulating layer of snow blocks around it. They came over to listen to the gramophone, and used his kitchen every Sunday for religious services, which they led themselves. They

FIGURE 4.4. Qikiqtat, August 1897. Inuit dwellings and drying arctic char are visible in the background. Nunavut Archives/William Wakeham/N-1983-002: 0024.

brought sealskins and other furs or ivory to the station year-round. Cardno had arrived in 1914 and planned to stay a year, but he was not picked up until 1917. After his supplies ran out, Inuit brought him food and took him ice fishing, and a woman showed him how to brew tea with local plants. Inuit suffered during these years from the lack of incoming trade goods. The late Elder Nowyook remembered how people coped without imports. They made ammunition out of what was available: melted lead for the bullets, tin cans or spent shells for the casings, matches for the firing caps, and cannon gunpowder that they crushed with a hammer to make a fine powder, mixing it with tiny pieces of wood if they were short of powder.[11] Malaya Akulukjuk, Qatsu's adopted sister, recalled sharing food with a Qallunaaq trader— possibly Cardno, as she says he was there for three years and it was during a war: "There was a white man at Kikiktaet [Qikiqtat]. We used to call him Mapatak. . . . He used to take care of me though he was a white man; he used to give me white man's food. I used to share my own meals with him when my foster mother [Aasivak] asked me to. He used to cry when I did that; he was very thankful. He didn't want to take our food, but we told him that he could eat with us."[12]

At the stations, Qallunaat and Inuit could develop relationships of trust and support. But many still lived quite separate lives. Cardno wrote about one of the winters he spent at Qikiqtat station: "Never was any man more thankful for a pack of cards. I sat alone night after night in the snow-covered station house playing away by the light of an oil lamp. The solitude was becoming a hardship. The Eskimos left me very much to myself. I was longing for conversation and my thoughts turned more and more to my folk at home. I started speaking to myself, a habit I found easy enough to get into but very hard to break."[13] Cardno was clearly not the focus of Inuit society at Qikiqtat. Much of what went on at the whaling stations was never recorded in Qallunaat records, often because Qallunaat did not know about it. In the three years that Cardno spent at Qikiqtat station, in the *ukiuq* evenings that he spent playing solitaire and missing his own family, what was going on in Inuit homes?

In the mid-1980s, Etooangat and Qatsu returned to the site of Qikiqtat whaling station in their old age. They were part of a group of Elders, researchers, and interpreters documenting the site's history as a commercial whaling station. Together they answered questions about whale hunting and processing, and about the archaeological sites, but in most of their memories, Qallunaat moved only on the fringes. Etooangat recalled sitting beside Qatsu at the station about seventy-five years previously, when she was a little older than him, a young girl and not married yet. They, and other Elders of their generation, remembered being children at the whaling stations, playing games that helped them develop necessary skills. They hunted lemmings and small birds.[14] They played a tag-like game called "Wolf" and another game called "Shipwreck" that involved team members trying to keep their balance while the other team ran at them.[15] While Cardno's recorded winter memories of Cumberland Sound rested largely inside the wooden station house, during *ukiuq* Inuit children played, ate, and slept alongside their parents and siblings in *qammat* dwellings. These were small homes—most being a single open room—that encouraged close family relationships and keeping peace with others. The walls were framed with whale ribs or wood, covered with two layers of sealskins, and insulated with a thick layer of heather that was replaced each year. The roof had windows made of translucent pieces of dried and scraped bearded seal intestine, sewn together with caribou sinew. The sinew expanded when wet and made the windows completely watertight. *Qammat* were built in a range of sizes, limited by each family's ability to heat their living space.[16] Inuit also lived in *tupiit* or sealskin tents in the warmer months, and they sometimes stayed in igloos or *igluvigait* on the sea ice or while traveling. But according to Etooangat, "everyone had *qammat* which they really called home."¹⁷ There was evidence of the commercial whaling trade inside these homes: families had metal cooking pots, and some installed wooden doors or flooring; others owned china figurines, accordions, and clocks. Yet a *qammaq* could be completely constructed from local materials. Heating and cooking were done with stone lamps, fueled by blubber, using wicks made from willow pods and moss. Warm, soft bedding made of caribou skins was piled on the sleeping platform. After fifty years of close contact with Qallunaat these were still very much Inuit homes, where Inuit relied on nothing imported for their basic shelter.

At Qikiqtat in 1984, Qatsu asked to see the site of her old *qammaq*. This was probably the family *qammaq* that had been haunted by the dead cooper in 1905, but Qatsu had later lived there alone, after her parents had moved into a bigger *qammaq*. It was unusual for young women—or indeed anyone—to sleep apart from their families. Qatsu recalled that her mother had always been protective of her, to the point of making her play inside at night instead of running around with the other children. She had always told her not to get too close to men. Then suddenly, when she reached marriageable age, her mother asked her to marry a man named Evic. Qatsu refused.¹⁸ When Qatsu would not be convinced, her mother moved her into the family's old *qammaq*, so Evic could "catch her." Clearly, however, Qatsu retained a degree of choice over whether to be "caught." She spent three years in the *qammaq* before finally agreeing to marry Evic. She was probably living there when Cardno was at the station. In retrospect, Qatsu concluded, Evic was a good hunter and a good provider. They had a good marriage. She joked that she had been scared of men until she married Evic, and then ever since he died she had gone back to being afraid of men again. "Even now!" she laughed, in her eighties.

Qatsu was audibly moved by the return to the site of her old *qammaq*. The interpreter explained that she was back at her childhood home and managing not to cry. It had been a long time since she had been to Qikiqtat, and she had given up trying to come here because she was so old. She was enjoying herself, and the weather was good.¹⁹ More recently, Inuit Elder Uqsuralik Ottokie recorded a similar reflection on the region where she spent her childhood, near the present-day settlement of Kinngait or Cape Dorset: "You may see other places but you still find the area where you spent your childhood days to be the best, most scenic and enjoyable place. Every time I go there it is very joyous. I always go back to our old camp to fish. When we start heading home I always think maybe this is the last time I will be visiting this place."²⁰

These emotions are hardly surprising—many people around the world are nostalgic about their childhood homes. But it is, I think, worth stressing

FIGURE 4.5. This print by Malaya Akulukjuk shows Qallunaat and Inuit homes, presumably at Qikiqtat, and Inuit butchering and sharing seals. I don't know what the artist intended, but I like how the footprints lead between Inuit and Qallunaat dwellings, but only partway. Malaya Akulukjuk, *The Hardship of Winter*, 1983, stencil on paper. Collection of the Winnipeg Art Gallery, Gift of Indian and Northern Affairs Canada, G-89-1361. Printmaker Jacoposie Tigilik, photographer Leif Norman.

how these sites, which often appeared so grand and desolate to outsiders, are also homes full of memories. Even Inuit who did not grow up at the whaling stations can still relate to them as home places. Rosee Veevee, Etooangat's daughter born well after the end of commercial whaling, was on a steering committee for the historic site at Qikiqtat. She recognized the extant play structures that children had made from rocks, because when she was growing up, she and her friends "would use old food containers from the Qallunaat," instead of stones, "[to] make . . . the same thing."[21]

Inuit altered their seasonal round to spend more time at the whaling stations, but they continued to hunt for personal use, to sew clothing and tents, to get married, to raise children, and to visit family and friends. Most of their daily activities, while often incorporating new imported equipment, were fundamentally the same activities their ancestors had performed. An Inuit family from 1805 entering Qatsu's family *qammaq* in 1905 would have been surprised by some things, like the abundance of wood and metal, the nearby station house, and the number of Inuit living on a small island for most of the year. They would, however, have had no trouble situating themselves in the *qammaq* and making themselves at home.

Upingaksaaq (Early Spring)

In *upingaksaaq,* Inuit often moved away from their *qammat* at the whaling stations into igloos or *igluvigait* on the sea ice. It was a welcome time of year. The days became longer, and game became easier to obtain. Migratory birds began to return, and the first seals were born around early March. In 1904, however, a devastating event occurred. Several Inuit families were asleep in their snow houses near Uttuusivik when a storm arose in the middle of the night. A woman awoke to two sounds: her kettle swinging back and forth, and shattering ice. The sea ice beneath their homes was breaking up.

Realizing the danger, the woman called out to the others to put on their clothing. One snow house split down the middle; the floor cracked open to reveal the black water below. Families ran out into the blinding snowstorm, trying to reach land, jumping from floe to floe. Out of the group of thirty-eight people, three died of frostbite and exposure when they became separated from the others or fell into the sea. One girl of about four was saved by her grandfather Kooto, who wrapped her in a caribou skin and carried her on his back until they were rescued by people from Uummannarjuaq whaling station two days later. Evie Nuijaut, one of the survivors, wrote up an account of the event:

> When the ice was breaking up, we fled our snow-houses, we were separated in the darkness—one party from the other. We could not see a sign of land anywhere on account of the darkness. When I left my house I took nothing with me, and had to leave even some of my clothing behind. Now when it began to dawn, we saw a small land to which we fled passing over broken masses of ice. When we reached the side of the land the waves and large blocks of ice were driven up on the beach. As we tried to get on the land on the cakes of moving ice, my father-in-law fell into the sea, but as I was close I laid hold of him, and helped him out. We got on the land which was only a very small uninhabited island. Here we remained for two days. We were very thirsty as we had but little water; we had something to eat as a seal was caught by one of the men. . . .

Up to here, Nuijaut's account described how, when disaster struck, her party drew on existing skills and knowledge, and on Inuit values of cooperation and resourcefulness. She continued: "I was not in much fear as we passed on over the blocks of ice for I was thinking of God, and I prayed much to Him. I feel thankful that I was taken in safety to land, for Jesus delivered me. While I passed from one pile of ice to another I wished to be guided to a place of safety by the spirit of God."[22]

Three years earlier, Nuijaut had become one of the first Christian converts in Cumberland Sound. Conversion was a transformative and disruptive process. Inuit debated whether local spirits, forces, and shamans should be superseded by a Christian God and a religion full of potent yet foreign agricultural symbols—bread and wine; shepherds and flocks. By the end of the whaling period, nearly all Cumberland Sound Inuit had integrated Christianity into their sense of home and of being Inuit. For many, the Christian faith would sustain them not only through local hazards like dangerous weather and ice conditions but also through challenges that arrived from outside.

Anglican missionaries arrived at Uummannarjuaq whaling station in 1894. Their first church, made of sealskins, was eaten by hungry dogs. It was seven years before they performed their first baptisms in 1901; Evie Nuijaut was among this initial group.[23] She later wrote her account of the storm for the missionary Edmund Peck, known to Inuit as Uqammak. He praised her writing as showing "how the power of God sustained and kept her in her time of peril."[24] Nuijaut would almost certainly have agreed with Peck's assessment, but while the God she called on was new, her type of prayer was not. Pleas formulated when trapped on ice floes or falling into the sea run far back into Inuit culture. Prior to accepting Christianity, Inuit called on helping spirits, or *tuurngait,* to assist them. As anthropologist Christopher Trott argues, Nuijaut and others overlaid Christianity onto their existing strategies for navigating difficult and dangerous situations.[25]

One of the first converts at Qikiqtat whaling station, Aasivak, also turned to Christianity when she faced a typically Inuit problem: a missing dog team. Aasivak had grown up next to the Anglican mission, where at a young age she was receptive to Christian teachings. She reportedly wanted to convert, but her mother forbade it. In 1899 she moved across Cumberland Sound to Qikiqtat to get married. A year or two later, she was out traveling when the men in her group spotted some caribou tracks. They left her with the dog team, which then ran away. Aasivak sobbed silently; she did not want her baby daughter, who was on her back, to hear her cries. A good dog team was an invaluable asset, difficult to replace and retrain. She wondered how she could get the dogs to return. She knew the shamanic teachings, but this time she thought about the new religion. Aasivak "trusted that somewhere there was a greater person than the people on earth." The dogs came back.

Aasivak later told this story to Qatsu Evic, the baby she had been carrying on her back that day. Qatsu wrote that, "Ever since that time, [my mother] started believing in God, even though she didn't want to turn to God because nobody had done that before."[26] Aasivak learned to read the Bible in Inuktitut syllabics and became very knowledgeable about Christianity.[27] She also collected caribou skins and cut out patterns for a full suit of caribou clothing, at least twice the size of what a normal Inuk would wear. The clothes were to be a farewell gift for Sedna, the sea woman who presided over the ocean and marine mammals. Several women sewed the suit together; then they threw it into the sea at Qikiqtat.[28] The women had evidently not ceased to believe in Sedna. The laborious act of making a full suit of clothing acknowledged her existence and power. Sedna was still a part of their homeland, but they were turning away from her to God, who, they were assured, watched over Cumberland Sound as well as everywhere else on earth.

Aasivak seems to have experienced a strong desire to convert, but for many Inuit this decision was heavy with fear and uncertainty. Shamans proved unable to halt the spread of new diseases and the decline of commercial whaling, but they were still powerful. Belief could be a life and death issue. Under shamanism, one person's transgressions could justify retributions from the animals, land, sea, or weather that would make the entire community suffer. After Aasivak threw the clothes into the sea, some Inuit feared retribution and returned to the old system of rules surrounding food and behavior. For a time, people at Qikiqtat separated into two separate camps of adherents to Christianity and shamanism, apparently not wanting to live with those who did not follow their rules. Daisy Dialla, a former Inuktitut teacher who worked to collect and record old stories, heard that when part of the population converted at Qikiqtat, at least three shamans remained active for a time. They "kept an eye on what [was] happening at Qikiqtat to make sure that what they're converting into [was] a good thing," until eventually they saw a light shining over the missionary Uqammak and accepted that his message was good.[29]

Christianity did not just divide the community of Qikiqtat; for a time it cleaved Aasivak's home in two. In early 1902, her husband Angmarlik claimed to have a revelation from Sedna. He preached a new religion and attracted followers. His teachings foregrounded helping the less fortunate, a core value of both Inuit culture and Christianity. He abolished many of the shamanic customs and restrictions on behavior, but retained the practice of spousal exchange. It is unclear which came first: Angmarlik's revelation, or his wife throwing the Sedna clothing into the sea.[30] But it seems likely that what the missionaries recorded as a syncretic revival was also, in part, a domestic dispute. It was a debate not just over the place of Christianity in Inuit society but over relationships between Inuit men and women. Daisy Dialla heard from Aasivak that her husband was resistant to Christianity because he "wanted to stay with the old ways, and that was because he could have any woman he wanted any time he wanted, that was part of the old ways." Aasivak said that she refused to sleep with Angmarlik until he became a Christian. She divided her household, living apart from him in a separate room attached to the family

FIGURE 4.6. Angmarlik in Pangnirtung, 1946. George Hunter/National Film Board of Canada/Library and Archives Canada/PA-166470.

qammaq. Angmarlik eventually gave in and converted to Christianity along with the rest of Qikiqtat.[31]

While Angmarlik was preaching in *upingaksaaq* of 1902, Inuit across Cumberland Sound at Uummannarjuaq held a series of meetings to which the missionaries were not invited. They decided, as a community, to embrace Anglican Christianity. Chris Trott notes that based on available evidence, it seems to have been a consensus decision. While the missionaries viewed conversion as deeply personal, Inuit at Uummannarjuaq saw it as affecting the safety and future of their entire group. Most converts took their new religion seriously. They brought their Gospels along when they whaled at the floe edge and generally abstained from work on the Sabbath even during whaling season. Inuit lay preachers went on to spread Christianity across a wide region.[32]

The process of shifting from one belief system to another extended far beyond the moment of official conversion. This can be glimpsed through Etooangat Aksayuk's stories of his early childhood. He was born at Qikiqtat on the cusp of Christianity. His birthdate is officially set in 1901, although Qatsu Evic's memories place it perhaps as late as 1905.[33] Aasivak was the midwife.

FIGURE 4.7. Koodloo Pitseolak at Uummannarjuaq, carrying a baby identified as Makikiuq. McCord Museum, MP-0000.598.43.

She may already have been a Christian, but a shaman still held Etooangat at his birth, because his older brothers and sisters "kept dying," and the shaman wanted to protect him.[34] Disease was a terrible concern. Eighteen Inuit had died of an epidemic at Uummannarjuaq in 1899, and twelve adults perished at Qikiqtat a year later out of a total population of about 140.[35]

By the time Etooangat was old enough to remember, he was being raised Christian, and most of the shamans practiced in secret.[36] He "became a young man when [Inuit] were still able to tell the old stories," although not as openly as before. His father Angutiqjuaq used to walk "back and forth . . . muttering strange things." He never explained this behavior to his son, but Etooangat later realized that his father had been speaking polar bear language, and that he must have been a shaman with a polar bear *tuurngaq* or helping spirit. Etooangat's mother Sulugaalik continued to sing the old shaman songs, rather than hymns, when she was alone. Another woman of Etooangat's generation, Koodloo Pitseolak, said that her parents used to go for walks away from the mission station to tell the old legends.[37] Etooangat also related that Kopanoa, one of the shamans at Qikiqtat, would predict when ships would arrive. "She would not actually see the ship," explained Etooangat, "but she could see it through her big light." This special light, *qaumaniq*, allowed shamans to see supernatural phenomena, as well as standard objects or events that were out of sight. Today some Elders use the word *qaumaniq* to refer to, among other things, the aura of Christ.[38]

Etooangat grew up when Inuit were changing the way they thought and behaved. They taught their children to be Christians and kept secret much of the knowledge of shamanism. Some of the spirits that they had seen or heard or felt out on the land—that had been a part of their sense of home—became invisible or unnamable to their descendants. Their homeland was overlaid with Christian prayers, stories, and experiences, like Aasivak's conversion and Nuijaut's escape over the ice. Christian markers of time like weeks, and celebrations like Christmas—Quviasuvvik or "the happy time"—became integral parts of the Inuit calendar. Today, both shamanism and Christianity are a part of Inuit cultural heritage, but the shamanistic tradition is not the central force it once was, especially in the cradle of Inuit Christianity that is Cumberland Sound.[39] Some old shamanic practices and stories continue to be circulated, but many have gone to the grave. Still, based on what I have seen, many of the old ways—particularly relating to how Inuit interact with and understand their homeland, the faith that many people place in external forces, and the priority on consensus decisions to shape a common future— have survived.

Upingaaq (Spring)

Upingaaq—the season when the snow melted, the ice broke up, most birds returned, and flowers bloomed—was the main bowhead whaling season. In late March or early April, when it was still *upingaksaaq*, but young seals were getting too wary and fast to be caught in their dens, the men on whaleboat crews got ready. Outfitted with new sealskin clothing made by the women, they headed out to the floe edge, where the land-fast ice met open water.[40]

The late Elder Pauloosie Veevee had heard about one successful hunt of an extremely large whale, led by the whaleboat captain Unirsagaaq.[41] Whaleboats were by this time equipped with harpoon guns and explosive darts. Everyone on the boat had a defined role. One crewmember shot the gun at the whale. The harpoon had a line attached, which played out so quickly that another man was stationed in the bow of the boat to pour water over the friction points. At the stern, others guided the rope and, if needed, spliced on extra lengths of rope from the neighboring whaleboats.[42] In this particular hunt, Unirsagaaq and his crew fastened onto the whale near the floe edge. The whale dove under the ice, stretching the rope, perhaps smashing the boat briefly up onto the floe.[43] "Sharks started attacking [the whale] under the ice," Pauloosie recounted. The whalers "could see globs of fat coming up from under the ice." This whale died under the ice, and they laboriously hauled it back up.

The dangerous part of the hunt was over, but the work was not. Sometimes the men spent days without sleep towing a dead whale directly into the station, with each stroke of their oars barely passing the previous one, because of the huge weight.[44] Other times they butchered it on the ice and transported it by dog team on huge sleds made specifically for floe-edge whaling. When the meat was butchered on the ice, it used to arrive at the station in bags, with blood pooled at the bottom that became, Elder Elisapee Ishulutaq had heard from her mother, "thick and very delicious."[45]

This whale was probably towed directly into Qikiqtat. It had died close to an island about fifteen miles away from Qikiqtat whaling station. The island has two names, which speak to Qallunaat and Inuit ideas of home. Inuit call it Uummannaq, meaning "shaped like a heart," a form familiar to them as hunters. Qallunaat whalers, using agricultural imagery, referred to it as Haystack

FIGURE 4.8. Unirsagaaq and Keenainak, whaleboat captains at Qikiqtat during the commercial whaling era, taken September 1923. Photograph Courtesy of the University of Alberta Archives, Accession #79-21-33-111.

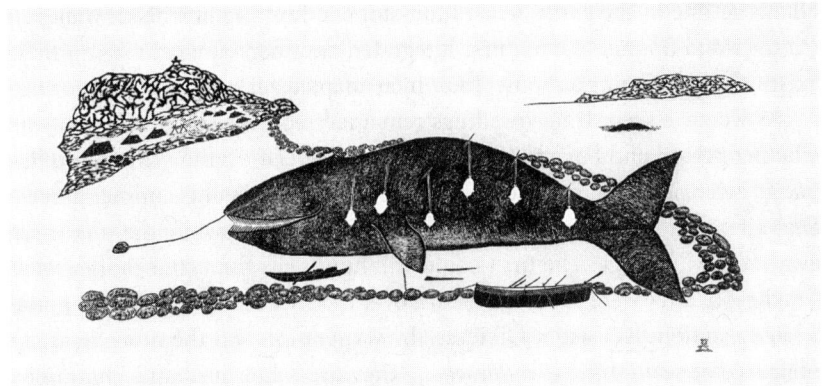

FIGURE 4.9. This 1984 print shows Qikiqtat whaling station when a whale was brought in. Josephee Kilabuk, *Qikittat*, stencil on paper. Collection of the Winnipeg Art Gallery, Gift of Indian and Northern Affairs Canada, G-89-1383. Printmaker Solomon Karpik, photographer Leif Norman.

Island.[46] Qatsu Evic recalled the arrival at Qikiqtat of an enormous whale caught by Unirsagaaq, larger than most whales, "so huge that my father and others looked as if they were children beside it."[47] The carcass would be left on the shoreline and every low tide, people would go down and work on it. It would take two days just to chop up the tongue, which was larger than a beluga whale.[48] The baleen came out easily as soon as the whale began to rot. Parts of the carcass would be eaten by Inuit, fed to dog teams, and cached for later use. Markosie Pitseolak of Uummannarjuaq station recalled that whenever a whale was killed, the successful whalers could also have all the Qallunaat food that they wanted, which normally was strictly rationed. "Sometimes we would even throw biscuits at each other like snowball fighting, thinking about how hard it was to get this food," he said.[49] Nowyook, a young man at Uummannarjuaq station, said decades later that he had never seen people happier than when a whale was brought in.[50]

Floe-edge commercial whaling became a core part of the Inuit seasonal round in the mid-nineteenth century. Men would hear a year in advance that they had been chosen to work on a whaleboat, and at least some fathers brought their sons along to teach them about whaling.[51] By the early twentieth century, however, bowhead whaling was winding down. New technologies and women's fashions left little demand for baleen, prices fell, and the fishery became unprofitable. A pound of baleen that fetched five dollars in 1900 sold for ten cents in 1912. By 1914, trading companies had taken over the whaling stations, and they dealt only opportunistically in bowhead whale products.[52] There were very few whales left anyway; in some years none were caught at all. Sometime in the early 1920s, Inuit stopped hunting bowhead whales in Cumberland Sound. Having first integrated commercial whaling into their existing ways of being at home, Inuit then adapted again to its loss.

As we have seen, Inuit dwellings remained recognizable throughout the whaling period and beyond. Yet during commercial whaling, Inuit families had to be creative and flexible to continue living in the same summer homes. Sealskins were at their prime for *tupiit* (summer tents) while the men were away at the floe edge. The men could still hunt seals there, but as they were "on the job," they were expected to turn over their sealskins and blubber to the whaling station. At least at Qikiqtat, the women took on the more typically male role of seal hunting in *upingaaq*. They used their husbands' harpoons while the men were at the floe edge. They could keep their catch.[53]

As with housing and clothing materials, Inuit selectively incorporated Qallunaat food but did not allow it to displace marine mammals as the mainstay of their diet. The men at the floe edge were allotted two or three meals a day from company rations. They would put aside part of each day's supplies for their families, presumably by supplementing their own meals with seal meat or by eating less themselves. They would only return to the station when the ice broke up, or if they brought in a whale, so they could not personally deliver this food to their loved ones. Instead they sent it back every two weeks with a dog-driver bringing sealskins back to the station. Elders Daisy Dialla and Inuusiq Nashalik had both heard that before syllabic writing was adopted, Inuit would tie pieces of string to the bundles of food, and the pattern of knots in the string indicated which family it was intended for. Daisy had heard that Inuit also coded gifts with items that recalled people's names. For example, the name Nettiapik means baby seal, so her items might arrive with a baby sealskin attached to them. The packages would have included seal meat, which did not interest the whaling companies, and imported food, especially hardtack biscuits. "Imagine that," smiled Andrew Dialla when he translated this. "Getting biscuits from the floe edge!"[54]

After the whaling stations closed and the bowhead whales nearly disappeared, the Inuit homeland changed. Inuit deepened another whaling tradition: hunting belugas, small white whales that live in Arctic and sub-Arctic waters. Inuit and Qallunaat had held commercial beluga drives at least as far back as the 1870s to supplement bowhead hunts, but the drives reached a peak in the 1920s and 1930s under the encouragement of the Hudson's Bay Company. Inuit used all available boats—many of which were old bowhead whaleboats—to encircle belugas and drive them into a bay where they would be beached as the tide went out. "It was a great sight and smell too," commented a police officer who observed the drive in 1926.[55] The hunters would

FIGURE 4.10. Taking whaleboats to the floe edge, early 1900s. These are the sleds and dog teams that would carry back sealskins and food from the men. This photograph was likely taken by the missionary Julian Bilby at Uummannarjuaq. Am,A49.206, © The Trustees of the British Museum.

wade around the bay in tall waterproof boots and kill the whales quickly with a shot to the head. They would have to be careful with their shots, Inuusiq Nashalik recalled. "There would be people and whales all over the place."[56] The next several days would be spent bringing the whales back to the post. Fur trade post records note as many as eight hundred belugas caught in a single drive, with 5,100 whales caught in total between 1923 and 1940.[57]

Although beluga herding required different knowledge and techniques than hunting large bowheads, the influence of bowhead whaling traditions and equipment was clear. Angmarlik, previously the Inuk manager of bowhead whaling at Qikiqtat, also coordinated the beluga drives.[58] Nearly everyone in Cumberland Sound would gather together for the beluga herding event; they would know to come to the trading post in Pangnirtung at the highest and fastest tide in July. Many people traveled in the old whaleboats they had received as payment for successful bowhead hunts, although some families acquired new boats with motors in this period.[59] Elisapee Ishulutaq, born after bowhead whaling ended, remembered crossing Cumberland Sound in a whaleboat, with the man in the stern setting the rhythm of the oars with his voice. She also had strong memories of being at home on quiet nights, when it was dark and a rowboat was approaching. She wouldn't be able to see

FIGURE 4.11. Beluga processing in Pangnirtung, 1929. Women are removing blubber after the whale carcasses have been towed into town. L. D. Livingstone/Canada, Dept. of Indian and Northern Affairs/Library and Archives Canada/e008440809.

the boat, but she could hear the oars dipping into the water as the boat came close to shore.[60]

The processing of belugas similarly recalled days at the bowhead whaling stations. Evie Anilniliak remembered how families slept in sealskin tents and spent two or three weeks hauling and processing the whales. The path leading from the shore to the blubber station became slippery with oil. Inuit ate as much of the outer skin as they wanted, and they cached meat for dog food, but they rendered the blubber into oil and packed it in barrels for shipping. The women carefully scraped the hard cartilage between the skin and blubber, which was then rolled up and salted in huge half-beluga pieces. When Inuit caught belugas for personal use, they used this cartilage, *majja*, to make ropes and the tips of dog whips.[61] As during bowhead whaling, women worked at the station, processing the blubber, while men killed the whales and hauled the carcasses. Social events were also similar. Previously men from Qikiqtat and Uummannarjuaq had met up and held friendly wrestling matches at the floe edge; they now began to do this during the beluga processing time.[62]

Beluga herding had something else in common with bowhead whaling: it was a colonial enterprise, in which Inuit provided very hard labor, and resources from their homeland were shipped out, with most of the profits and products ending up in Qallunaat hands. I wondered if herding whales in this intensive way was also a Qallunaat idea, but the late Inuusiq Nashalik did not think so. He thought it was probably the brainchild of the Inuit whaleboat captains, because they "knew how to hunt all the animals around here." They were the ones who "knew the easier way to get at all these animals."[63] In his opinion, the Qallunaat knowledge of the land was too sparse to come up with an idea like that. Jamesie Mike had heard that the first time beluga herding was done in Cumberland Sound, it was done by a group of women armed with stones. The men had left to go looking for caribou and were surprised to return to a camp full of food. The women knew that the water was shallow there, and they figured out that they could drive the beluga far inland and beach them during the highest tides. This legend is reflected in the name of the bay, Millurialik, which means the place where stones are thrown.[64] In beluga herding and bowhead whaling, imported technologies like motorized boats and guns were used to great effect, and existing skills were adapted to new economic activities. But the success of the hunts rested on ingenuity, flexibility, and the knowledge that came from being at home in this place.

Aujaq (Summer)

During the days of commercial bowhead whaling, most Inuit went caribou hunting around August, in *aujaq*. Caribou coats harvested at this time of year made first-rate winter clothing and bedding. Only a few Inuit would stay behind at the whaling stations to unload and reload the incoming ships, clean up the stations, and prepare them for the winter. Many or most family groups headed to the vicinity of Nettilling Lake. Located inland northwest of Cumberland Sound, it is the largest lake in the Canadian Arctic Archipelago and covers over two thousand square miles.

In 1899, Qatsu Evic's grandparents disappeared on a Nettilling Lake hunt. This happened just before Qatsu was born, when her mother Aasivak was newly married and pregnant. Aasivak accompanied her parents, Aullaqiaq and Itutaluk, on the first part of the journey up Cumberland Sound, then said goodbye. They never came back.

When Qatsu was young, every year her family would follow the same route her grandparents had taken. Her mother wanted to find some trace of them. Their bodies were never located, but their *qamutiit* sleds were later discovered, facing—in Qatsu's translated words—"towards home." Home, in this sense, meant Cumberland Sound, and probably specifically Uummannarjuaq whaling station, where her grandparents had lived before their disappearance.[65]

Qatsu and her family do not seem to have considered Nettilling Lake "home," not in the same core sense as the coastal areas where they spent most of their time. Travel inland could be particularly dangerous, as people were relatively isolated from other families and from rich marine resources. Qatsu later recalled that the only time she had truly experienced hunger was once "up in the area of [Nettilling] Lake where there [are] hardly any other sources of food besides the caribou. The sea is better," she added, "as there [is] food to be found almost anywhere."[66] There are other stories of Inuit failing to return from the Nettilling Lake region. Decades later, the late Joanasie Dialla looked every year for another family that had disappeared. Finally, next to a river, he found skeletons of a woman holding a baby. He thought they must have been trying to cross when everyone drowned, except the woman with the baby on her back. She had managed to get to the riverbank, but had died there.[67]

These stories highlight risks associated with travel in the Arctic, although not all risks were truly local. A missionary speculated that Qatsu's grandparents had died not directly from some stereotypical Arctic hazard like cold or starvation but from the influenza epidemic raging at the time.[68] And unlike most Qallunaat narratives of Arctic perils, Inuit stories of Nettilling Lake contain a strong element of continuity: people searched year after year for missing loved ones, continued to hunt caribou there, returned over and over again to the same place. The full route up to Nettilling Lake requires four portages.[69] In 1910, the German explorer Bernhard Hantzsch noted that while most routes he traveled on with Inuit were indiscernible to him, even he could follow the Nettilling Lake trail at the first portage. It had been so well traveled that a beaten-down track was clearly visible. At the lake, he recorded his Inuit traveling companions "running hither and yon, viewing the old tent sites used from time immemorial."[70]

Nettilling Lake was far from a terrifying "empty wasteland." It was still a part of Inuit homelands, a known place that provided resources necessary for a good life on the coast. Inuit women turned caribou skins from Nettilling Lake into warm clothing and bedding, making their families comfortable through the winter, and equipping their husbands to safely hunt seals in frigid temperatures. Feeling at home often relies in part on practices, traditions, or exchanges that require distant connections or travel. Although Inuit life for most of the year was focused around the sea, the *aujaq* caribou hunt was key to a sense of home.

The annual caribou hunt was a great point of continuity before, during, and for decades after the whaling period. When the foreign whalers first arrived, Inuit were traveling to caribou country on large boats made out of bearded seal or walrus skins. By the end they made the trip in wooden whaleboats. The hunt was necessary enough that Inuit interrupted whaling season for it. They were not allowed to take the companies' regular whaleboats on the trip, and instead used boats they had acquired outright over generations of commercial whaling.[71]

Although not everyone went to Nettilling Lake every year, Inuit who rou-

INUIT IN CUMBERLAND SOUND   133

tinely headed there would store pieces of wood or whale ribs along the trail to
help them slide their boats.[72] Dogs would carry packs.[73] Usually the men would
leave the main family camp, roaming for days or weeks to accumulate enough
caribou. If all went well, they would end up with a huge pile of skins. They
would pack all of this into one tight bundle, which also contained sinew and
fat and dried meat, and which they waterproofed by wrapping it in two more
caribou skins. One man would carry it down to the water.[74] In 1883, Franz Boas
described Inuit returning contentedly from the hunt in mid-October:

> Many of [the groups] brought whaling boats which they had obtained from
> a whaling vessel. They were filled to the gunwales with skins obtained during
> the summer. Men, women, and children laughed, sang, and talked in the boats.
> The dogs howled, often one or the other went to the kettle filled with chow that
> was standing in the middle of the boat. The man steering the boat sat alone on
> an elevated seat, looking serious and majestic while navigating his boat. If the
> wind was unfavorable, it became necessary to row. If a seal popped its head
> out of the water and there was no particular reason to hurry along, they would
> stop and every gun was ready to greet the seal should it come up again for air.[75]

When Inuit returned to the whaling station, the women would break the bale
of skins apart and begin to work on clothing after the first sea ice was forming.
It was taboo to work on them before that.[76]

The caribou caught each year provided meat, marrow, and fat; their sinew
was used for thread, bowstrings, and bow backing; and their bones and antlers for numerous tools and adornments and games. Some of these uses were
supplanted by imported technology during the whaling period, but there was
no imported material that compared to caribou skins for winter clothing. Missionaries, whalers, anthropologists, and explorers all adopted it too, if they
were going to spend any time in the region. Each adult needed six or seven
skins a year for their outfits. Skins were also prized for bedding. If the annual caribou hunt failed, it could compromise a group's ability to provide for
itself.[77] Winter hunting consisted largely of waiting at seal breathing holes out
on the sea ice, and it was not safe to do that without proper warm clothing.

Qallunaat also valued caribou, but their ideas about how to relate to caribou did not always accord with Inuit ones. The *aujaq* hunt continued throughout the whaling period, but Etooangat recalled that Inuit whalers were not
supposed to hunt caribou at all for the rest of the year, as sealskins and blubber
had more market value. Restrictions on caribou hunting later become more
stringent and official, as the Canadian government feared the caribou population was declining and imposed wildlife management policies on Inuit from
the 1920s onward. In the 1940s, Canada condemned the number of caribou

kills at Nettilling Lake and officials told some Inuit to remain on the coast. This likely included Qatsu Evic, who mentioned that she and her husband once moved temporarily to Nettilling in order to find a good winter dog-team trail in, but Qallunaat told them to leave. Inhibiting Inuit from traveling to Nettilling—which was not only critical for obtaining caribou skins but was also a meeting place that drew Inuit from all over South Baffin Island—struck at the core of Inuit society.[78]

Even after the Canadian government pressured Inuit to move into the year-round settlement of Pangnirtung in the 1960s, many families continued the *aujaq* caribou hunt at Nettilling. Most eventually stopped for a variety of reasons. Parents put their children in school, sometimes under threat of losing family allowance payments, which represented a major cash source for many Inuit and were supposed to be offered to all Canadian families without restrictions. Many sled dogs, which had carried saddlebags on long hunts, were shot. Some adults obtained jobs in town and could not get the time away. Substitute materials—often inferior—for winter clothing and bedding were available. New federal government restrictions were placed on caribou hunting.[79]

Still, when they could, Inuit kept hunting caribou, passing on knowledge about caribou, eating caribou, and making clothing and boots incorporating caribou fur. They kept traveling to Nettilling, including by snowmobile. They hunted caribou in different places and in different seasons. Inuit hunters know their homeland: they are adept at recognizing and remembering other areas that have been disturbed by grazing caribou as well as areas that have plants that caribou like to eat.[80] The late hunter Pauloosie Angmarlik recalled being up near Aujuittualuk (the Penny Ice Cap) in the summer and recognizing winter caribou droppings. He knew then that he should return that winter, and he found caribou then.[81]

Inuusiq Nashalik, who was born at the very end of the whaling days, clearly remembered following his father on the annual caribou hunts. Sometimes they would be out for a month, with just the clothes on their backs, their knife, and their gun. They would live off what they caught. Inuusiq's father would make him walk such a long way that sometimes he would cry, hoping his father would stop and rest. Later he realized that his father had been effectively training him to hike long distances, giving him the skills necessary to provide for his own family when he grew up, teaching him to be at home in this place. Inuusiq, when in his nineties, shot a caribou from a boat in *aujaq* of 2008. Referencing the bundle of skins that a man from each group would carry down the trail after the annual hunt, and the caribou that had fed and clothed him throughout his long life, Inuusiq commented that he "would not be able to carry all the caribou he's caught over the years."[82]

At the time of writing, aerial surveys indicate that the caribou population on Baffin Island has decreased dramatically, as it has in many other locations. The territorial government has initiated strict hunting quotas. Caribou management remains emotional and controversial for many people who live with caribou across North America, not just in Cumberland Sound. In my experience, Inuit are still very much thinking about, talking about, and relating to caribou. Caribou remain important even when they are not being seen or hunted.[83]

Ukiaksaaq (Early Fall)

In September 1922, the schooner *Easonian* burned to the ground at Qikiqtat.[84] The ship had already broken down multiple times. Because of a defective engine, the crew had navigated mostly by sail from Scotland, at some points resorting to "the back-breaking expedient" of towing the schooner with a boat. They had to stop in Greenland for repairs. After finally arriving in Cumberland Sound, the captain and crew were out beluga herding with local Inuit when the engine's flywheel flew off and smashed. It was decided to beach the ship, remove the propeller, and sail back to Scotland. When the crew reached Qikiqtat, Inuit began towing the *Easonian* into shore. Suddenly a fire started in the engine room and the fuel tanks exploded. No one could save the ship. Clustered outside the burned-out wreck, all ten sailors were left "literally in the clothes they stood up in." The Inuit at Qikiqtat welcomed the shipwrecked men into their *qammat* dwellings. Women measured them for winter clothing.[85] The season of *ukiaksaaq* was coming, with its snows and night frosts.[86] In Cumberland Sound, this was the time of year when if a ship had not yet arrived, it would not be coming.

The captain of the *Easonian*, John Taylor of Dundee, was well known in Cumberland Sound, and in many ways he was at home there. He was experienced and capable, having sailed on Arctic whaling ships since he was a boy.[87] The previous year he had arrived after a whale was towed into Uummannarjuaq, and both Inuit and Qallunaat celebrated. The crew played music on the ship's gramophone, and the Inuit hosted a feast with caribou, seal, and walrus meat. In addition to whale oil and baleen, Inuit had saved seal, caribou, polar bear, and wolf skins to trade.[88]

In 1922, however, the fire consumed Taylor's supply of trade goods. Taylor and several Inuit men rowed a whaleboat up Cumberland Sound to look for another Scottish vessel, the *Albert*. One night soon after, an Inuk inside a *qammaq* at Qikiqtat commented, "Ship coming." Suddenly the dogs began to bark. People ran outside and saw the lights of the *Albert* approaching. The

FIGURE 4.12. Captain John Taylor. Courtesy of Andrew Dialla.

Easonian's captain and crew sailed home. Sail was the appropriate term, since the *Albert*'s engine failed ten hours into the trip, and it took them forty-three days to reach Scotland.[89] John Taylor would never return to Cumberland Sound, but he was not forgotten there.

The burning of the *Easonian* is one of several possible places to mark the end of commercial bowhead whaling. The ship's owner, Robert Kinnes and Sons, opted not to replace it. They pulled out of Arctic trading. As Etooangat recalled, "after Taylor . . . nobody came back, and there were just Inuit people living here [at Qikiqtat]."[90] Bowheads were hunted sporadically for a few years following Taylor's last voyage, but whaling was basically over.[91] Inuit now lived and traveled differently in their homeland. Families continued to disperse from the old whaling stations, spreading out to *nunaliinnut* or *ilagiit nunagivaktanginnut*, much smaller communities that are often roughly and inappropriately translated in English as "outpost camps." In the mid-1950s, grass began to grow on the trampled areas of Qikiqtat again.[92]

The *Easonian*'s owners sold their holdings to the Hudson's Bay Company for £2000, helping the HBC establish a monopoly in Cumberland Sound. The transition from whaling stations to the HBC was not smooth. Most Inuit never became the dedicated fox trappers the HBC had hoped for, even when

FIGURE 4.13. The Hudson's Bay Company post at Pangnirtung, ca. 1925. McCord Museum, MP-0000.598.200.

fox prices were high. Inuit focused on seal hunting to provide for their families; they trapped only when they wanted to buy something from the store: tobacco, ammunition, a boat. In other words, most Inuit were not trapping to accumulate profits. They were trapping just enough to get the imported goods that had become part of their sense of home during the commercial whaling era. The culture of the HBC was also different from that of the whaling companies. Inuit whalers had received basic rations even if their hunts were unsuccessful, and they continued to hunt for the whaling station even if the Qallunaat ran out of supplies. The HBC, in contrast, introduced a system wherein "everything had a price."[93] At least officially, Inuit were not entitled to trade goods unless they brought furs to the post—no matter how hungry they were, how few foxes there were that year, or how much food was stockpiled in the warehouse. The new system did not account for the lean times and hard years that Inuit knew were part of living off the land. Nor did it fit with Inuit practices of sharing food, even with outsiders, in times of need.[94] Reflecting the profiteering and self-interest with which white people have often become associated in Canada's North, Elder Pauloosie Veevee deemed these early HBC employees "the perfect Qallunaat."[95]

The earlier relationships with commercial whalers were portrayed as more positive overall, at least in stories I have heard and read. Daisy Dialla told me, "There's never any stories of unfriendly Qallunaat [whalers], like when we go south [today] we'll pass a lot of Qallunaat and none of them will look at us, there are no stories like that."[96] Many Qallunaat whalers were, or had once been, family. In 1909, there were about 168 Inuit living around

Uummannarjuaq station. Nineteen had Qallunaat fathers, and presumably many others had whaling grandfathers or great-grandfathers.[97] While not all the relationships that produced these children would have been long-term or even consensual, the children intertwined whalers and whaling history permanently into their families and communities. Separation from whalers could be marked by sadness and longing. Elder Jamesie Mike, whose grandfather was a Qallunaat whaling captain, spoke of how Inuit used to *unga-*, or long for, departed whaling relatives and partners.[98]

John Taylor is among the whalers with descendants in Cumberland Sound. On an early visit to Cumberland Sound, he began a relationship with an Inuk woman named Arnaqoq. Arnaqoq and other women used seal oil as a beauty product, working it into their hair. Daisy Dialla had heard that Arnaqoq attracted many suitors, but her father only approved of Taylor.[99]

Arnaqoq and John Taylor had a son who they named Joanasie. This is the Inuit pronunciation of John, but Joanasie's descendants have heard that he was not named after John Taylor. He received the name from an Inuk, John Aggaarjuk, who accompanied the German explorer Bernhard Hantzsch on an expedition to the west side of Baffin Island in 1912 and died not long afterward. Aggaarjuk's young daughter, Koodloo Pitseolak, would call the baby Joanasie *ataata*, or father.

Joanasie was born in 1913, just after a ship was wrecked, when people were fishing tobacco and other goods out of the wreckage. The ship was probably the *Ernest William*, which was beached at Qikiqtat on September 4 of that year. Shipwrecks could be a boon to Inuit. Wrecks literally became part of Inuit homes, as Inuit salvaged wood for their dwelling frames and floors. Inuit also used the wood for tools, and they removed the metal, which they would sometimes melt down for bullets or make into harpoon tips. This shipwreck also became a *niriujaq*, or omen, in Joanasie's life: whenever he dreamed of people hooking tobacco out of a wreck, he would catch a seal the next day. Daisy Dialla, Joanasie's daughter, noted that the wreck is still at Qikiqtat.[100]

John Taylor acknowledged Joanasie as his son. On subsequent voyages he brought food, clothing, and a gramophone for his child. He also brought candy, but Joanasie was not used to sweets and didn't like it. John Taylor may also have given his son an accordion, because when the HBC store began stocking the instruments, Joanasie bought one and to his family's surprise, he knew how to play it right away.[101]

In many ways Joanasie had a difficult life. His father John Taylor left for the last time in 1922, and his mother Arnaqoq died not long after. Effectively an orphan, Joanasie was adopted by two uncles. He later married and had twelve children of his own. The eldest, David, died of illness before the age of two

FIGURE 4.14. Pangnirtung in July 1951, soon after the Dialla family moved there. Qaqqaqtunaaq, the woman walking away from the camera, is Andrew Dialla's namesake. She used to warn Andrew's father of the dangers of hunting at polynyas, but she stopped after he returned from a polynya with a huge male seal, at a time when they had no blubber in camp. Wilfred Doucette/National Film Board of Canada/Library and Archives Canada/PA-166461. Information on Qaqqaqtunaaq from Daisy Dialla.

and is buried at Naujajaapvik, one of five different Inuit settlements the family lived in before relocating to Pangnirtung in the late 1940s. In Pangnirtung, Joanasie worked as a special constable for the RCMP. This was a tough job, where Inuit employees were expected to bridge the cultural divide but could often feel trapped within it. When Joanasie began working for the police, most Inuit still lived out on the land. Pangnirtung was home to the HBC store, the mission station, the hospital, and the police station. It was not prime seal hunting territory. Daisy Dialla remembers her mother's reaction—"Eeeee!," the Inuktitut version of "Ewwww!"—when Joanasie told the family they were going to move to Pangnirtung. Joanasie traveled often for police work. He patrolled not only Cumberland Sound but also across the Cumberland Peninsula to Qivittuuq (Kivitoo) and occasionally as far as Iqaluit. Sometimes he traveled with Etooangat, who was working for the doctor. Daisy recalled the sound of her father's dog team returning home: "There would be at first no sound at all coming from the land, [then] our local dogs would start howling, that means there is a dog team coming . . . even before the people knew, the dogs knew, so they would start howling. It was great when the dogs started howling because that means Dad is coming back."[102]

Joanasie's children grew up knowing that their grandfather was a whaling captain. At the end of the 1960s, when the Canadian government was assigning surnames to Inuit, Aasivak said that Joanasie's father's name was "Tialla"—an Inuktitut pronunciation of "Taylor." The name was written down as "Dialla," and Joanasie and his children took that surname. Joanasie's son Andrew Dialla began searching for information on his Scottish relatives. Working from his home in Pangnirtung, Andrew did the research to track his family down. Decades later he found them. In 2007, one of John Taylor's Qallunaat grandsons, John McGuinn, visited Pangnirtung to a warm and generous welcome at the airport packed largely with his relatives. The event was

FIGURE 4.15. John McGuinn and Andrew Dialla, grandsons of Captain John Taylor, in Pangnirtung in 2007. Courtesy of Andrew Dialla.

filmed for the Canadian television program *Ancestors in the Attic*. During the show, Andrew addressed John by the English kinship term "cousin," thanked him sincerely for coming, and spoke about the importance of family, saying that Inuit rely on their families for comfort, companionship, and support.[103]

Inuit have a long history of migration, and it is not uncommon for branches of families to be separated for long periods. When Inuit from different communities meet today, they often try to find out how they are connected to each other. Family ties are a large part of feeling at home: they are important for security and comfort but also for keeping alive a sense of history and possibility. In finding his Scottish relatives and welcoming them into his extended family, Andrew reestablished ties that had connected Cumberland Sound with Dundee for generations but that had been dormant for almost a century.

Ukiaq (Fall)

Ukiaq, the fall freeze-up period, was the hardest time of the year for Inuit in Cumberland Sound during the commercial whaling era. Both Uummannarjuaq and Qikiqtat whaling stations were on islands. They were conveniently located for whaling from about April through October, but during freeze-up they were surrounded by unstable forming ice, which prevented hunters from moving around in search of game. They were not places Inuit would have chosen for a fall camp. *Ukiaq* was often a season of hunger at the stations.

Etooangat told a story about a famine in *ukiaq* to Margaret Nakashuk in 1994.[104] He explained that "in the late fall and early winter there used to be a lot of blizzards and gales" at Qikiqtat whaling station. Inuit didn't always have a lot of meat during freeze-up, but they would receive tea and other goods from the whalers. They would also try to hunt birds and collect clams and seaweed around the island.[105] One fall, the ice did not become safe for travel until after Christmas. Etooangat called this "a dark time," which it literally was. Without blubber to burn in their oil lamps, Inuit had neither heat nor light in their homes. Etooangat lamented that there had been lots of blubber that summer, but it had been taken away by the supply ship months ago. He believed that if the blubber had been in the storehouse, "they might have been able to access a little bit of that." His translated words suggest that these Qallunaat were not exactly generous with stockpiled supplies, but he did not place blame.

Etooangat instead explained the strategies his community used to find food when they could not hunt. They ate bowhead whale tongue, which takes a long time to rot: "During that summer, before the famine in the fall, they caught four whales at the same time, and it was at that time that a lot of

FIGURE 4.16. Men dragging seals, probably near Uummannarjuaq, in the early 1900s. Am,A49.111, © The Trustees of the British Museum.

tongues were available. They knew the tongues would last a long time and they were saving them for dog food." He described how the famine ended: "But a favorable wind finally blew from the northwest . . . and a big huge chunk of ice that had formed over there beached at Qikiqtat, a big huge slab of new ice. All the men got their harpoon ready and their paddle." The interpreter Andrew Dialla explained that the paddle was a small piece of wood the hunters could attach to their harpoon. If a hunter shot a seal, he would chop off a piece of the ice slab and paddle an "ice boat" out to retrieve the seal. It is not safe to take boats out in the late fall because ice can form around the boat.[106] In this situation, Inuit used new guns and wooden and metal tools, combined with older Inuit hunting techniques, to end the hunger. Etooangat concluded: "And they were catching seals then and a lot of them were dragging seals back and that was a very happy time that I remember . . . that was the one extreme famine that I remember when I was a child."

Hunger features prominently in many Qallunaat accounts of Inuit life. Qallunaat tell of Inuit in Cumberland Sound coming to the whaling ships in desperate search of food, dogs crawling about and dying, hunters having no luck for weeks, entire families so hungry they ate their skin clothing. In 1847 at Naujaaqtalik, 160 Inuit had gathered to trade with Qallunaat whalers, and twenty of them reportedly starved to death after the ships left in the fall.[107]

These deaths and hardships must first be understood in their context of commercial whaling. Whaling ships—especially overwintering ones—provided important hunting gear and rations, and dead whales provided food for Inuit and their dog teams. But at the same time, the local population increased exponentially with transient Qallunaat workers and new Inuit migrants; new illnesses kept hunters at home more often; whaling stations were not well located for year-round hunting; and some winters were particularly difficult after Qallunaat ships had sunk or failed to arrive. One letter, from the Anglican minister Julian Bilby in 1905, accuses Qallunaat whalers of appropriating Inuit sealskins and blubber and sometimes meat even in times of shortage. Bilby also claims the whalers provided Inuit with too little ammunition, to keep them tied to the station. He writes: "It is no uncommon sight in Blacklead [Uummannarjuaq] to see a whole family lying in bed all day long in the winter to keep warm because they have no oil for their lamps, and no meat to eat. Their clothes are too thin for attempting to hunt, while there are large hogsheads of this oil alongside the tents which the traders are waiting to ship to Scotland, and in their storehouse are hundreds of seals while the Eskimo sometimes are starving in their tents and no help given."[108] Bilby's letter was written in the context of poor relations between missionaries and whalers; other sources record Inuit keeping blubber for themselves as well as traders redistributing blubber or meat as needed over the winter.[109] But accounts like Bilby's show that at least some food shortages were directly caused or exacerbated by Qallunaat. These could be the same Qallunaat who wrote as if they were bearing witness to Inuit in their "natural" state of hardship.

Qallunaat often fixated on Inuit hunger. Yet the Inuit stories I have heard suggest that rather than being an ever-present issue at the whaling stations, hunger was seasonal. Indeed, if we look at the timing of the Qallunaat accounts cited above, the most severe cases of hunger all took place during *ukiaq*, with the possible exception of the starvation at Naujaaqtalik, which we only know took place sometime after the whaling ships left. Elder Etooangat Aksayuk did not remember hunger as a problem for most of the year at Qikiqtat whaling station. He said, "Every year whenever the ice freezes over, any kind of famine or hunger would end. From the ice freezing until the late summer, there would be no hunger. Only in the late fall would there be hunger times."

Inuit stories of hunger are also, in my experience, presented in a very different way from most of those written by Qallunaat. First, I have heard very few Inuit stories of the whaling days that mention hunger. There could be many reasons for this, including not wanting to discuss secondhand knowledge, or to share these stories with me, or to pass them on at all. But Etooangat portrayed hunger as an unwelcome but cyclical event at Qikiqtat whaling station. It was not a normal or constant state but an infrequent challenge that Inuit were capable of handling together. In another telling of the same story, Etooangat stressed that "when hunters caught their seals the women would distribute portions to those families that didn't have any," explaining that the whole community knew each other and knew who was in need. He described how the late fall could be difficult at Qikiqtat but then enumerated all the ways Inuit tried to procure food. He stated that he did not remember anyone dying of starvation.[110] Even in stories of hardship, there is a stark contrast between Inuit who speak about the Arctic as home and Qallunaat writing about it as a harsh and unforgiving place.

In terms of overall nutrition, Daisy Dialla speculated that Inuit were probably better off after whaling ended, because they could be much more mobile. They lived in smaller communities and moved freely to seasonal hunting areas.[111] The overall population of Cumberland Sound seems to have increased by the 1930s.[112] Some people still went hungry at times, and at least one group starved when they became isolated and could not find animals, but others did not experience hunger. Saullu Nakashuk recalled in 2008 that she never went hungry as a child after whaling because "there were animals all around" at her community of Qimiqsuut, and her father was a "very active hunter and a successful hunter, and she grew up always having fresh good food."[113]

However, the decades that followed commercial whaling often evoked a different kind of hunger—craving for food that was no longer available. Many recent Elders remembered people who had longed for bowhead products. Pauloosie Veevee's grandmother had died yearning for bowhead whale meat, wanting a taste of the food she had remembered as a child and that had always been part of her culture.[114] Etooangat also mentioned this craving, and Elisapee Ishulutaq recalled old people being nostalgic for *mattak,* the outer layer of skin. She commented, "it would be their absolute favorite yummy food," along with the cartilage from the whale's throat.[115] "Our mothers and our fathers craved bowhead *mattak,*" recalled Saullu Nakashuk, adding that she had longed to find out what it tasted like.[116] Bowhead whales had provided other products too, for which Inuit now had to devise alternatives. Inuusiq Nashalik had seen his father make the bases of sled runners out of bowhead jawbones, which, when polished, glided extremely well. Once jawbones were no longer available, Inuit began using a type of moss to make the sleds glide, which worked in very cold weather but wore out quickly late in the season.[117]

Without bowhead whale products, the Inuit homeland was a changed place: one that Inuit adapted to, but not without longing and a sense of loss.

In 1994, near the community of Igloolik west of Baffin Island, a respected Elder named Piugaattuk announced as he was dying that he wanted to eat *mattak* again. Inuit hunters killed a bowhead whale and were charged under Canada's Fisheries Act, although the charges were later dropped. This was the first of several hunts that led to an established legal bowhead whale hunt in Nunavut today.[118] Many Inuit and conservation organizations have since agreed that the Davis Strait–Baffin Bay bowhead whale population is on the rise, if still below precommercial whaling levels.[119] Industrial whaling is still banned, but Inuit in Nunavut territory are now collectively allowed five bowhead whale hunts per year, and the meat and *mattak* (skin) are distributed by plane around the territory. Pangnirtung hunters towed a whale into Qikiqtat whaling station in 1998 and again in 2013 and 2016. The 1998 hunt gave many Elders a chance to try *mattak* for the first time. Pauloosie Veevee was not so impressed. He thought his piece of *mattak* tasted like an "old rubber boot." He didn't care much for it, "but if he was very hungry, it would be a different story." Jamesie Mike agreed that the *mattak* from the top part of the head supposedly does taste like a rubber boot, and he wondered if this was the piece that Pauloosie got. He very much enjoyed his piece.[120]

*

Inuit did not hunt bowhead whales for most of the twentieth century. Elders portrayed this to me as a difficult but short gap in Cumberland Sound history: Inuit were whalers before they met Qallunaat, they were whalers with the Qallunaat, and now they are whalers again. Through the twentieth century, Inuit anticipated and prepared for whaling's eventual return. They continued to tell stories of whaling, to travel on the land, to hunt and maintain relationships with seals and caribou and other animals, and to pass knowledge onto their children, despite facing many new obstacles introduced by Qallunaat. North of Cumberland Sound in the hamlet of Qikiqtarjuaq, near where Etooangat's father was born, the hunter, guide, and outfitter Billy Arnaquq recently described how integral seasonal changes remain to Inuit life:

> Elders have a passion for spring, because the winter is so long and cold. People always have a passion for when the spring is coming, the birds are coming, the animals are more plentiful. . . . It is like that season comes for you to enjoy it. It is inside the people. Some men have a passion for early fall, when the harvest is plentiful and animals are migrating through. Some have a passion to go out seal hunting in the winter. Some of them cannot wait until the ocean is frozen. . . . Once you are part of that, it just becomes part of you. [121]

FIGURE 4.17. The artist Lipa Pitsiulak's vision of a qalupalik. Lipa Pitsiulak, *Qalupaliq*, 1979, stencil on paper. Government of Nunavut Fine Art Collection, on loan to Winnipeg Art Gallery, 979.77.1. Photographer Leif Norman.

Inuit today remain tied to local ecological cycles, as well as to imported notions of time like Sundays, the workweek, Christmas, and the school year. This continued connection to the land—which people have fought and sacrificed to maintain—has arguably been key to maintaining their sense of home. So, I think, was an understanding that the world is neither fully knowable nor always what it seems.

Near the end of Etooangat's life he recorded a few of the "old stories," the ones he had heard as a child that dated from before Inuit conversions to Christianity. One of these stories begins: "My old late mother once ate *qalupalik*."[122] It is an attention-grabbing opening line. *Qalupaliit* are walrus-sized sea monsters, sometimes considered to be part human, that lurk under the ice or by the shoreline. They are feathered, or wear clothing made of eider duck skin. They steal children by stuffing them into pouches on their backs, the way

Inuit mothers carry their babies. *Qalupaliit* have two flippers: one is pointed and can make a shrill sound that paralyzes anyone who hears it; the other is bulbous, deformed, and "globby . . . like Jell-O." As they move around under the water, they make a low beeping sound.[123]

A hunter had once returned to his mother's camp bearing chunks of whale blubber and skin. He claimed to have found a dead beluga. The story seemed plausible, since the beluga skin was scarred and full of holes, as if seagulls had been picking at the floating carcass. But the man had rubbed the skin over rocks to rough it up. The meat was not beluga; it was *qalupalik*.

It was impossible to kill a *qalupalik* in its regular form, Etooangat explained. When faced with one, a hunter would ask it to change shape. The hunter would say, for example, "*nettiuniakutik*," please turn into a ringed seal. This hunter had asked the monster to become a beluga, and it did. "That's how my mother ended up eating a *qalupalik*," Etooangat explained.

"In the old days," Etooangat continued, "when they saw a *qalupalik* they would very slyly go after that *qalupalik* by pretending not to go after it." As they threw the harpoon, they would make their request, and when it died, "it would emerge from the water again as whatever you wanted it to turn into." Etooangat gave an old word for this shape-shifting, *pilutitaminik*.

There are Inuit alive today who have returned from trips out on the land, shaken up, having heard the sound of a *qalupalik*.[124] Etooangat concluded his section on *qalupaliit* by saying that he had never seen or heard one of the monsters. "Even though I went alone all over the place," he said, "I've never seen anything like that." But he also said, "From what I've been able to ascertain, *qalupaliit* probably exist."

When some Inuit talk about *qalupaliit* as if they could exist or have existed, it is not evidence of gullibility, but rather the opposite. In my opinion, it implies that they know their homeland better than anyone else, and that old knowledge specific to their land is still valuable and does not always fit easily into Qallunaat scientific categories. They remain aware that there is a lot they cannot be certain of, and that it is best to constantly plan for a range of eventualities. In the words of English writer Robert Macfarlane, "the true mark of long acquaintance with a single place is a readiness to accept uncertainty: a contentment with the knowledge that you must not seek complete knowledge."[125] Out in a small camp or cabin in the vastness of Cumberland Sound, with the sounds of shifting ice and the immensity of the water, it is much harder for anyone to state categorically that such creatures could not exist. I think this relationship with the land, this belief that shapeshifting is in the natural order of things, and this openness to possibility and change has been a source of strength for Inuit, in a home that has in many ways been closed in around them. It is also worth remembering that the homeland of *qalupaliit* was the world that Qallunaat whalers overwintered in, whether or not they were aware of it. It was a place where things were not always what they seemed, where people could see beyond the horizon and move back and forth between the human, animal, and supernatural worlds. It was a place where a whale might not even be a whale.

EPILOGUE

At Home

Inuit in Cumberland Sound lived through great changes related to commercial whaling, but these were in many ways eclipsed by what came later. In the 1950s, the Canadian government began taking a more active approach to administering the Arctic, developing its resources, and shaping the lives of its citizens. In the 1960s, they pressured Cumberland Sound Inuit to resettle into the community of Pangnirtung. This is where the recent stories in this book were recorded or documented, and where nearly all Inuit in Cumberland Sound live today. The Qikiqtani Inuit Association recently summarized some effects of these relocations, which occurred across Baffin Island and beyond:

> By 1975, all but a few Inuit families lived in government-created permanent settlements, and many of them felt that their lives had become worse, not better. The decision to give up the traditional way of life was almost never an easy one, and once made, it proved to be irreversible. Inuit made enormous sacrifices by moving into settlements, living in permanent housing, giving up their qimmiit [sled dogs], sending their children to school, or accepting wage employment. Once they had made their decision, they discovered that government assurances of a sufficient number of jobs and better living conditions were illusory in many cases. Looking around, Inuit often felt and saw despair as they, their family members, and their neighbors struggled to adjust to circumstances beyond their control, even though some received benefits from living in settlements, such as less risk in daily life, better health care, and options to work for wages rather than hunt. Settlement life often imposed a new form of poverty, and hindered access to the land and the country food that nourished them.[1]

Different concepts of home were and are at stake in Canada's Arctic. Canadian government spending, while often inadequate, has historically centered on Qallunaat markers of a good home: permanent wooden houses, schools, health centers, hockey rinks, post offices, radio stations, and airports. For Inuit, accessing these amenities has entailed compromising other things—like local mobility, autonomy, language, opportunities to pass certain kinds of knowledge onto children, and a relationship with the land built through everyday interactions over a lifetime—that were integral to their own sense of home. There have been many consequences, for Inuit, of becoming part of a country where most citizens knew little about Inuit homelands and imagined the Arctic as alien and inhospitable.

The resettlement into Pangnirtung marked a break with the past, but Inuit did not abandon their previous home places. At first, many families returned to the site of their old communities for months each year to hunt. They referred to these trips as going back to their land or their home. It became harder to make these extended trips after seal pelt prices crashed in the late 1970s and early 1980s, following protests against seal hunting. But people in Pangnirtung know where their families used to live, and many still travel there when they can.[2]

Most Inuit in Nunavut today were born and raised in settlements like Pangnirtung, and have dreams or jobs or hobbies that would be recognizable across North America and beyond, even if they often lack equal means to achieve their goals. They continue to fight on many different fronts to uphold their ideas of home, demand social equity with other Canadians, and integrate certain southern concepts on Inuit terms. Some of this resistance is visible and public: court cases, protests, policy statements, community-based programs, press releases, media interviews, documentary films, and statements on social media. Just a few recent examples range from Alethea Arnaquq-Baril's film about the seal hunt, *Angry Inuk;* to the wellness programs of the Ilisaqsivik Society in Clyde River; to that same community's successful legal case against seismic testing that reached Canada's Supreme Court; to Inuit Tapiriit Kanatami's *National Inuit Suicide Prevention Strategy*; to the grassroots food security group Feeding My Family; to Tanya Tagaq's music and Twitter feed.[3] Yet resistance also takes place quietly, with a grandparent telling stories in a living room, families boating out to their cabins, a government employee going hunting on Saturday, mothers murmuring to their babies in Inuktitut and carrying them on their backs in *amautiit,* friends entering homes without knocking, and people posting on Facebook that seal meat or spaghetti is available at their house for anyone who is hungry. Inuit quietly but firmly claim their place with Qallunaat outsiders as well. In my own experience, individual Inuit have gently poked fun at my Qallunaat ways, turned some conversations toward serious stories of injustice, patiently encouraged attempts at Inuktitut,

EPILOGUE   151

corrected my assumptions at meetings, or been unavailable for meetings when hunting conditions are good. To me these are reminders that, as a Canadian government employee, I may be in my own country but I am in someone else's homeland, and I have much to learn.

What do I think I have learned so far? First, that while some Qallunaat are integrated and respected in Inuit communities and families, Qallunaat as a group need to work to change the ways we continue to relate to Inuit and Inuit lands. The stories in this book point to some of the stereotypes, misconceptions, hierarchies, and ways of seeing within Qallunaat society that have historically led Qallunaat to fail to understand the Arctic as a homeland. An awareness of history will not in itself lead to Qallunaat fully respecting Indigenous rights, or taking steps recommended by Indigenous experts to resolve everyday inequities, but I think it can help Qallunaat to recognize blind spots and alternative possibilities. Second, studying the commercial whaling period has impressed on me the wealth of strategies Inuit had for coping with, mitigating, and avoiding hardships; but also how colonialism—especially the resettlement into permanent communities—has made some of these strategies less workable. Third, while it is not the responsibility of Inuit to welcome Qallunaat or to change their behavior, I have felt in my heart the Inuit values of *inuuqatigiitsiarniq,* respecting and caring for others, and *tunnganarniq,* "fostering good spirits by being welcoming, warm, and inclusive."[4] There is no doubt that the litany of crises that define Nunavut in southern media coverage are real. But they also coexist with community feasts, and children skating free across the sea ice, and tea cooked over heather fires, and what may be the best party games on the planet. Above all, I have come to see that Cumberland Sound is a beautiful homeland, and worth fighting for.

The Inuit Elders I cite in this book speak to issues that resonate far beyond their homeland: they give specific examples of living with environmental change, of relating well to people and the land, of being patient and tolerant, and of viewing truths as relative rather than absolute. From my perspective, Inuit have a long history of offering support to others facing challenges, coupled with a reluctance to generalize or dictate how others should do things. I encourage readers to seek out Inuit voices and writings, to learn about Inuit knowledge in greater depth than I can know or present here. Inuit words, while often embedded in the land and individual experiences, have always reached beyond the local.[5]

*

One issue I have not dealt with enough in this book is how Inuit, even those who spent most of their lives in Cumberland Sound, had connections to other

FIGURE E.1. The Mike family in Pangnirtung in 1942. Jamesie Mike is the child wearing the cap; his father Mike or Kanajuq is on the far right. NWT Archives/Archibald Fleming fonds/N-1979–050: 0739.

places. They traveled and adopted ideas or goods they found useful from other Inuit and Qallunaat. Inuit were not just expert whalers or guides or interpreters, as Qallunaat often described them; they were also seasoned experts on dealing with Qallunaat. A story Jamesie Mike told me about his father in 2008 speaks to this broad experience and expertise. I asked Jamesie if he knew of any Inuit who had gone abroad on whaling ships, and he told me that his father, Mike or Kanajuq, moved across hundreds of kilometers of Inuit territory, and once went as far as Scotland.⁶ As Jamesie heard the story from his father, Kanajuq was born near what is now Taloyoak, in the Kitikmeot region of Nunavut. Kanajuq's father was a Qallunaaq whaling captain, and his mother was an Inuk. The captain acknowledged Kanajuq as a son, but when Kanajuq was still a child his Inuit family moved to Salliq (Southampton Island) in Hudson's Bay, almost five hundred miles away as the crow flies. I do not know if this journey was made by dog team or whaling ship, but recall Jamesie's comments in chapter 3 that Inuit would set out across similar distances just to go visiting. Long-distance travel did not begin with commercial whaling; it is an Inuit tradition and part of how Inuit make the Arctic home.⁷

As an adult, Kanajuq encountered a group of Inuit from Cumberland Sound who had traveled to Salliq with the whaler-trader William Sivutiksaq Duval. When they returned home, Kanajuq sailed with them. He and his son were the only ones from Salliq who came back with the ship. Kanajuq, a widower, reportedly went because of Kilabuk, one of the women on Duval's ship. Kanajuq married Kilabuk and settled in Cumberland Sound.⁸

In 1919, at the very end of commercial whaling, Kanajuq and his son Akpa-

EPILOGUE   153

lialuk traveled to Peterhead, Scotland.[9] Jamesie had heard that his father was a friend and guest of the captain—it made sense that the son of one whaling captain would have a friendly relationship with another. Jamesie had never heard whether his father fell ill overseas, but he was reportedly seasick on the Atlantic crossing. Kanajuq also "spoke about the yearning for Inuit food, really wanting to eat Inuit food over there and not being able to."

In Scotland, Kanajuq learned some English, and how to operate and repair coal-powered steam engines. After he came back, he maintained a small coal-powered boat that was used in beluga herding operations at Usualuk in Cumberland Sound. When the Hudson's Bay Company established their trading monopoly in the 1920s, they purchased and refitted the boat. Jamesie added, "[Then] something on it broke and they couldn't fix it, so they went and got my father to work on it again."

Kanajuq could fix the engine when the Qallunaat Hudson's Bay Company employees could not. Many Inuit today are experts at tweaking and fixing outboard motors, snowmobiles, and Coleman stoves; or at film-making, baking, working in government jobs, and many other imported tasks. This is not to say Inuit have uncritically adopted everything new. Throughout the commercial whaling era in Cumberland Sound, Inuit traded for guns and metal needles and wooden boats, but they retained unrivalled local technologies like kayaks, outdoor fur clothing, and dog teams. The shooting of many dogs in conjunction with the relocation to permanent settlements in the mid-twentieth century remains a heart-rending and watershed event in Inuit history.[10] Now that most Cumberland Sound Inuit live far from their family's preferred hunting areas, using and maintaining motorized transport has become central to maintaining a connection with the land—with the part of home that exists beyond the settlement. The literary critic Craig Womack has argued that the definition of "traditionalism" should include, "anything that is useful to [Indigenous] people in maintaining their values and worldviews, no matter how much it deviates from what people did one or two hundred years ago."[11]

For many Inuit and Qallunaat whose lives were shaped by cross-cultural encounters and far-flung travels, it seems that home was not simple to define or pin down. William Sterry, a successful retired American whaler, looked back on his time living with an Inuit family and longed for raw seal meat. Men like Sidney Budington and Ambrose Bates bemoaned the Arctic climate but chose to sail north over and over again. Hannah wrote to Sarah Budington, grieving and wishing she could visit the grave of her baby in America. Hannah's brother Inuluapik reportedly pleaded with whaling captains to take him abroad, but her husband Ipiirvik tried to dissuade an Inuit family from making the trip south. Kanajuq met his wife Kilabuk and moved to Cumberland

Sound because of a whaling-trading ship captained by William Sivutiksaq Duval, a German-American with an Inuit name and family and descendants. These stories do not just illuminate the complex nature of home and attachment. They show how making home can hinge on unlikely and meaningful connections across cultures, while at other times it involves ignoring or resisting other people and their ways of thinking.

*

In this book, I have also tried to show how the definition of a harsh environment is relative and contingent on many human factors. In the nineteenth century, popular American descriptions of the Arctic verged on the magical. It was a place where the sun did not always rise and set, where compasses could not be relied on, where strange lights danced in the winter sky and mirages appeared on the horizon, and where nature's laws did not seem apply. Yet to Inuit, the temperate world was equally disorienting, with its wider variety of stars, a relatively short twilight, a more closed-in landscape, different wind patterns, a denser population, and a focus on terrestrial rather than marine resources. Inuit visitors were surprised that people could live so far back from the sea and still obtain the necessities of life. They remarked on the vast gulf between rich and poor, and on the number of lights, the number of people, the large buildings, the farm animals, the "lights of the houses glittering like so many stars." One Inuk compared the crowds of people to mosquitoes.[12] When Uugaq from Cumberland Sound traveled to New York City, his reported reaction was, "God damn! Too much horse—too much house—too much white people. Women? Ah! Women great many—good!"[13]

Uugaq was lucky to return home safely; many Inuit who went abroad did not. Inuit experienced the United States as a dangerous place, and neither was it safe for many Americans. The era of American Arctic whaling and exploration was marked by lynchings, Indian wars, mine explosions, tenement and factory fires, high infant mortality, malnutrition, contagious disease, and great inequalities of wealth. It is bookended by the Civil War and the First World War.[14] This is not to say the nineteenth-century Arctic was an easy place to live either. Many Inuit and Qallunaat have expressed great admiration for the skills, understanding, technologies, and knowledge of *inummariit*, "real Inuit" who lived on the land.

Qallunaat suffering in the Arctic is legendary, but it often had less to do with the landscape and more to do with how Qallunaat traveled through it, where they chose to travel, and how they imagined the entire region as the antithesis of home. And the Qallunaat who documented Arctic peoples so extensively in the Victorian era were not simply observers of Inuit at home.

EPILOGUE

They were witnessing a society they had disrupted, and they often contributed, intentionally or not, to where and how Inuit lived and died. To sum up, it is not that the Arctic is any less of a home than temperate climates. It is that the Qallunaat vision of what a home is, and how to live there, has—for now—come to dominate how outsiders think and talk about the Arctic.

Taken together, these Inuit and American stories show how difficult it can be to feel at home in a new place, or to hold onto a sense of home in the face of drastic change. Living in new or changed surroundings can require learning to think, move, or react in new ways. It involves emotional, intellectual, and physical work. It often demands breaking and reorienting one's preconceptions of what is natural, of what is right and wrong. Many Americans and Inuit died longing for home, or trying desperately to reach it, even as they showed impressive creativity and strength in dealing with hardships. Today, when people are forcing or pressuring others around the world to leave their homes in unprecedented numbers, it is important to consider the tremendous effort and suffering inherent in these situations, and how they are shaped and complicated by factors such as power relationships, colonialism, or financial hardship. Leaving or remaking home can of course have positive outcomes, and some people long for this opportunity. But people should also be able to choose to stay in their home regions and to determine what that home looks like.[15] Inuit and American stories of home and homesickness offer a glimpse into everything people fight to keep—and everything that can be lost or taken away.

Appendix: Methodological Essay

I would like to briefly address some issues surrounding my research and Qallunaat research in the Arctic more generally. Although I cannot and should not offer definitive guidance, I think it is important for historians to reflect publicly on the uncertainties, challenges, and debts around our own work.

Inuit advice on many kinds of research in their homelands needs to be better heard and respected. This includes the basic question of when, how, or if such research is appropriate. I obtained a Nunavut Research Institute licence for this project, consent from interviewees with the understanding that I hoped to publish some of what they told me in this book, and permission from family members to use quotes from 1990s interviews with two deceased Elders. However, I acknowledge that this project falls short of models of collaborative cross-cultural research, and I do not claim that it is "community-based." Inuit mentors and friends provided invaluable advice, but the project was neither suggested nor fully shaped by people in Pangnirtung. If I continue to do oral history research, I will always be striving to improve at it through reflecting on my own missteps, on concerns that people involved in oral history projects have shared with me, and on thoughtful published critiques.[1] In my current job with Parks Canada I have been fortunate to work with local staff and committees of Elders, Youth, and Hunters and Trappers Association representatives, known as the Inuit Knowledge Working Groups for Auyuittuq National Park. Most recently we have collaborated on a place-names project that I believe is important to many local community members as well as to Parks Canada. Our consultation process is still evolving, but I can see the benefits and high quality of work that result from these kinds of ongoing relationships.

This study is based on a wide variety of Inuit and Qallunaat stories, and many of the Inuit stories were told in Inuktitut. Inuktitut contains countless words and concepts that are difficult or impossible to fully express in English, especially when people speak about the land. For example, in 2000 the North Baffin Elder George Agiaq Kappianaq tried to explain the behaviour of an *ijiraq*, a spiritual being that lives inland. He used the Inuktitut word *qajaaksaqtuq*, which he defined as "when the [sea] ice is thin and the waves make the ice ripple," to explain that "an *ijiraq* is able to make the land ripple."[2] These terms have no direct English translation, and full comprehension is contingent on familiarity with the Arctic landscape and its animals. Skilled interpreters — whose powerful role in daily life in the past and present Canadian Eastern Arctic is arguably understudied and underappreciated — can translate Inuktitut stories well, but some of the poetry and the layers of significance are lost.[3] My knowledge of Inuktitut is very far below grasping the complexities of Elders' language. I want to acknowledge this as an issue within this book, in part to give Qallunaat readers a sense of what is not here. On a basic level, I also note that I have kept to the translated wording except for changing some translated statements in the third person (e.g., His father told him . . .) to first person (My father told me . . .).

Much of what I came to understand about Inuit oral history I owe to Andrew Dialla, a talented interpreter from Cumberland Sound whose grandfather was a Scottish whaling captain. The Elders Andrew and I interviewed together in Pangnirtung in 2008 mostly told stories aimed at my own level of comprehension of Inuit culture, but when Andrew interpreted earlier recordings that had been intended for an Inuit audience, it took him longer to answer my questions than to translate the tape. He patiently explained to me where historic communities were located, how ringed seals were hunted in various seasons, and about Inuit supernatural beings and legends. He told me about the deceased Elders whose voices we were hearing and described some of the old words and narrative structures in their stories. He evocatively brought the world of whaling stations into the tiny windowless room in the Angmarlik Visitor Centre, where we sat with a tape player for many afternoons as the hillsides of Pangnirtung Fiord turned a brilliant autumn red outside. He also provided detailed information, collected over decades, about his grandfather John Irngutaq Taylor's story. Andrew generously took the time to read and offer important comments to some key draft sections of the manuscript. When I reference Andrew Dialla, as I frequently do in the chapters that deal with Inuit history, I am referencing his work as that of another historian.[4]

Still, I take full responsibility for any errors, and the opinions and analyses are my own unless otherwise stated. I am a Qallunaaq woman who grew up

APPENDIX   159

in southern Canada, and although I have been traveling to Pangnirtung since 2006 I have never lived there. Culturally my background is very different from both Inuit and American whalers in the nineteenth century, and from Inuit in Pangnirtung today. I have tried to listen and learn as much as I can, but I have no doubt that there are many subtleties and interpretations that I failed to hear or grasp in both Inuit and Qallunaat stories, and many stories that as an outsider I will never hear.

Despite these limitations, I still think it is critical that Qallunaat historians who are grappling with Arctic history take Inuit experiences seriously and work toward understanding them through the words and knowledge of Inuit. I also want to make a case for spending time in the places we study. Travel in Nunavut is expensive. Extended stays were possible for me by participating in and then working on a University of Manitoba field school (unfortunately no longer running at the time of publication), and later by the kindness of Panniqtuurmiut who let me housesit. The opportunity to stay in Pangnirtung for weeks or months during *upingaaq* and *aujaq*, and to go out on the land with Inuit families, forever changed the way I read Qallunaat records of this region, and the way I think about my own country and its history. Also, the beauty and weight of this land cannot be conveyed in words or pictures.

Trying to write about Inuit and Qallunaat experiences highlights the limits of knowing the past. The stories that survive today represent only a fraction of what occurred, only a fraction of what was thought. Those that do exist serve to highlight that everyone does not have equal means to pass on the experiences and memories they consider important. I draw on accounts of Inuit by Qallunaat like Charles Francis Hall and Adolphus Greely, because in many cases they are the only surviving accounts of what Inuit travelers experienced. These records are difficult to work with. They are often blatantly racist by today's standards, and it is usually impossible to know whether the statements about Inuit are accurate, or what was left out. To interpret these accounts and to try to better understand them, I have relied on conversations with Inuit and Qallunaat, and on recent writings including memoirs, ethnographies, and oral history projects, especially the series of interviews by Inuit students with Inuit Elders, published by Nunavut Arctic College. Both Inuit and non-Inuit experts are generally very careful not to extrapolate beyond their own time or place, and Inuit societies within living memory are not the same as those of 150 years ago. My interpretations about encounters between Americans and Inuit in the nineteenth century should not be seen as definitive. I try to make it clear in the book when I am conjecturing using more recent sources, which I cite in the footnotes. Readers can judge for themselves whether they agree with my methods or my interpretations. In the end, while I respect that some people may disagree with me, I felt that the Qallunaat stories about Inuit should not be allowed to stand at face value, but neither should they be excluded entirely.

A final note on spellings and place names: I have tried to follow the spelling conventions laid out by the Inuit Cultural Institute (ICI) in 1976, but I have retained some original spellings in quotes from other sources, and I have not standardized the spelling of most people's names. Across Nunavut, Inuit Heritage Trust is engaged in a huge project to document and share Inuktitut place names. Most of the thousands of place names they have collected are not yet official, but, wherever possible, I have sourced names from their draft maps available at ihti.ca. I use these Inuktitut toponyms out of respect for Inuit knowledge and with the assumption that they will mostly become official place names. Where Qallunaat assigned their own place name, I also give it in parentheses if known.[5] For present-day Inuit hamlets, I use the official name at the time of publication. Therefore, I refer to Qikiqtarjuaq not Broughton Island, but I use the anglicized name Pangnirtung. Community members voted to retain Pangnirtung as the official name in 2004, although some residents prefer the correct spelling Panniqtuuq, which means "place of bull caribou."

Notes

Prologue

1. On the *Polaris* crew's drift, see, e.g., Blake, ed., *Arctic Experiences*, 201 (darkness quote) and 365 (fearful perils quote); Hendrik, *Memoirs*, 65–73 (taking guns and ammunition on 69); United States Navy Department, *Annual Report;* Davis and Hall, *Narrative of the North Polar Expedition;* Bessels, "Smith Sound and Its Exploration," 385 (party becoming stranded).

2. For records of ice floe drifts in the late nineteenth and early twentieth centuries, see Gilder, "Dangers of the Ice Pack," 278–80; Hall, *Arctic Researches*, 245, 308, 555–56; Nowyook, interview with Marc Stevenson, G-1985-007-0010, NTA; Uvdloriaq, "The Narrative of Qitdlarssuaq," 16/1–20/1. The last source concerns the solitary drift of the hunter Piuaitsoq, which has the most in common with the *Polaris* drift as it lasted for months.

3. Inuusiq Nashalik, interview, August 27, 2008, AVC. Trans. Andrew Dialla. For more on my use of oral history, see the methodological essay in this volume.

4. This is of course only a small piece of Inuit knowledge related to traveling safely on the ice, and the translated quote does not reflect the large specialized vocabulary surrounding ice and ice travel. Other hunters have stressed the importance of bringing extra snow or ice for drinking water when hunting at the floe edge, and of watching currents and wind conditions to predict where a broken-off piece of ice might touch land again. As with most things to do with ice, the rules are contingent and evolving due to climate change. See Laidler, "Ice through Inuit Eyes," esp. 186, 191–93, 247, 253; Bravo, "Sea Ice Mapping," 173–74. On Inuit and sea ice, see also Gearheard et al., eds., *Meaning of Ice;* Inuit Cirumpolar Council, "The Sea Ice Is Our Highway"; Krupnik et al., eds., *SIKU;* Laidler, ed., *Inuit Siku (Sea Ice) Atlas,* sikuatlas.ca.

5. Hans Hendrik's account of the ice floe drift has some of the calmness of Inuusiq's accounts; for example, he describes the party as "not having suffered any real misfortune." Hendrik, *Memoirs,* 74. For more recent treatments of the drift see Loomis, *Weird and Tragic Shores;* Nickerson, *Midnight to the North;* Parry, *Trial by Ice;* Henderson, *Fatal North;* Heighton, *Afterlands.*

6. Hans Hendrik interpreted by Hannah, as cited in US Navy Department, *Annual Report,* 352, and Hendrik, *Memoirs,* 74. On Kudlago, see Harper, *In Those Days,* 57–60.

7. As reported in Hall, *Arctic Researches,* 470–71. This quote presumably fails to fully convey Budington's feelings, as he chose to make over a dozen Arctic voyages and spent much of his adult life in the North.

8. Andrew Dialla, pers. comm. See also Anawak, "The Environment," 10402; Ipellie, "Nunatsiaqmiut"; Inuit Tapiriit Kanatami, *Nunatsiaq*.

9. On the power of outsiders' ideas and stories about the Arctic, see Price, "The Arctic Is My Home"; Obed, "The True North"; Cameron, *Far Off Metal River*. For a few varied examples of the use of homeland, see Nuttall, *Arctic Homeland*; Inuit Tapiriit Kanatami, "Maps of Inuit Nunangat"; Whitney Lackenbauer, *Arctic Front, Arctic Homeland*; Canadian Museum of History, "Nunavut: The Inuit Homeland," http://www.historymuseum.ca/cmc/exhibitions/hist/frobisher/frhom01e.shtml; David Miller, "Canada Has a Critical Opportunity to Protect Healthy Arctic Homeland," March 30, 2015, http://blog.wwf.ca/blog/2015/03/30/canada-has-a-critical-opportunity-to-protect-healthy-arctic-homeland/; and Berger, *Northern Frontier, Northern Homeland*.

10. On increasingly somber depictions, see Peck, "Art of the Arctic"; Cavell, "Going Native," esp. 32–33; Loomis, "Arctic Sublime"; Riffenburgh, *Myth of the Explorer*. On the picturesque, see the work of Ian Maclaren (cricket pitch example in "Limits of the Picturesque," 107). On fewer depictions of Inuit, see David, *Arctic in the British Imagination*, 47–56. For examples of Arctic imagery, see Savours, *Search for the North West Passage* and Potter, *Arctic Spectacles*.

11. Although not the focus of this book, much academic scholarship skillfully examines the meanings, complexities, and diversity of Victorian Arctic exploration and its retellings and representations. See, e.g., Robinson, *Coldest Crucible*; Cameron, *Far off Metal River*; David, *Arctic in the British Imagination*; Riffenburgh, *Myth of the Explorer*; Bloom, *Gender on Ice*; Potter, *Arctic Spectacles*; Spufford, *I May Be Some Time*; Cavell, *Tracing the Connected Narrative* and "Publishing Sir John Franklin's Fate"; Savours, *Search for the North West Passage*; McCorristine, "Searching for Franklin"; Lewis-Jones, "Heroism Displayed"; Davis-Fisch, *Loss and Cultural Remains*; Garrison, "Virtual Reality"; Hill, *White Horizon*; King and Lidchi, eds., *Imaging the Arctic*; Maclaren, "Aesthetic Map," and "Retaining Captaincy"; Belyea, "Captain Franklin"; Craciun, *Writing Arctic Disaster*. See Coates and Morrison, "The New North," 640–41, on the enduring popularity of exploration histories. For recent retellings of Arctic disasters, see many of the books on the list of Arctic literature reviewed by the *New York Times* in Wilson, *Spiritual History of Ice*, 221–22.

12. I use the word Inuktitut since the speakers I cite are from Baffin Island. The term Inuktut is increasingly used, especially in Nunavut, to collectively designate all regional forms of the Inuit language.

13. Martin, Rak, and Dunning note in Freeman, *Life among the Qallunaat*, xiii: "The etymology of the term 'qallunaat' is subject to debate, but it is used variously to refer to 'Southerners,' 'white people,' or even 'English speakers.'" Freeman considers the word's roots in *Life among the Qallunaat*, 86–87, and notes possible connotations of greed and materialism. She also comments that the word's roots do not refer to skin color, saying "Inuit too are aware of different races, but not of colour. I have never known or been taught to identify a different race by skin colour." On "whiteness" as a shifting category, see, e.g., Jacobson, *Whiteness of a Different Color*; Ignatiev, *How the Irish Became White*. It is possible that the term Qallunaat also had different meanings in the nineteenth century. The singular form is Qallunaaq, and there is also a dual form, Qallunaak.

14. Andrew Dialla in Inuusiq Nashalik, interview, September 15, 2008, AVC.

15. On the multifaceted violence of American colonialism and imperialism in this period see, e.g., Blackhawk, *Violence over the Land*; Jacoby, *Shadows at Dawn*; Kaplan, *Anarchy of Empire*.

16. Simpson, *Mohawk Interruptus*, ix.

17. On the Arctic as a mythical place, see Kollin, *Nature's State*; McGhee, *Last Imaginary Place*; Grace, *Canada and the Idea of North*. On "imaginative geographies" constructed by out-

siders, see Said, *Orientalism*; Clayton, *Islands of Truth*; Campbell, *In Darkest Alaska*; Gregory, "Imaginative Geographies."

18. The full text of the Nunavut Agreement is online at nlca.tunngavik.com.

19. Kumlien, "Fragmentary Notes," 11 (English names). The two Inuktitut names are used in Pangnirtung today, but others are used elsewhere. Pauloosie Veevee and Billy Etooangat, pers. comm.

20. Friesen and Arnold, "The Timing of the Thule Migration."

21. M'Donald, *Narrative*, esp. 3, 74–77.

22. Ross, "Annual Catch"; Goldring, "Southeast Baffin Island," 1.21. Ross notes that his estimates do not include mortally wounded whales, so the total number killed is likely significantly higher.

23. On resource extraction in Canada's North, see, e.g., Piper, *Industrial Transformation*; Cameron, *Far Off Metal River*; Keeling and Sandlos, eds., *Mining and Communities*.

24. On Hudson Strait, see Barr, "The Eighteenth-Century Trade." For more on Cumberland Sound commercial whaling, see Ross, *Distant and Unsurveyed*; Goldring, "Inuit Economic Responses," "Whaling-Era Toponymy," and "Southeast Baffin Island"; Stevenson, *Cultural Persistence*; Eber, *When the Whalers*.

25. Harper, "The Collaboration"; Müller-Wille, ed., *Franz Boas* and *Inuit and Whalers*; Boas, "Arctic Expedition"; Knötsch, "Franz Boas' Research Trip."

26. See Qikiqtani Inuit Association, *Community Histories: Pangnirtung* and *Achieving Saimaqatigiingniq*.

27. Tuan, "Home," 164.

28. On geographies of home, see, e.g., Blunt and Dowling, *Home*; Briganti and Mezei, eds., *Domestic Space Reader*; Brickell, "'Mapping' and 'Doing'"; Sara Ahmed, "Home and Away"; Ahmed et al., *Uprootings/Regroundings*; Porteous and Smith, *Domicide*; Nowicki, "Rethinking Domicide." See also Davidson, Bondi, and Smith, eds., *Emotional Geographies*; Smith et al., eds., *Emotion, Place, and Culture*; Erickson, "Preface."

29. Klinkenborg, "The Definition of Home," *Smithsonian Magazine*, May 2012, http://www.smithsonianmag.com.

30. On *angirrattinni*, see Michel Kupaaq in Saladin d'Anglure, "La toponymie réligieuse," 116–17. For contemporary uses of *angirraq*, see, e.g., "The Nunavut Hansard Inuktitut-English Parallel Corpus" and Spalding, *A Multi-Dialectical Outline Dictionary*, both available through inuktitutcomputing.ca. Other words related to home in Spalding's dictionary include *sungiunaktuq*: "this which causes one to feel at home or at ease: a person one knows well, a place or country which one is used to or has inhabited for long time; a custom one has long practiced or [is] familiar with"; *tungasuktuq*: "he feels at home; feels free in or at home with his surroundings"; *ungajuq*: "he longs for . . . or yearns for . . . persons or living things"; *nuna*: "country; homeland" (*nuna* is an important word with many other connotations); *kajjarnaktuq*: "that which causes homesickness or longing for familiar things." This term *kajjaanaqtuq* means much more than just "that which causes homesickness" on South Baffin Island; see, e.g., Rasmussen and Akulukjuk, "My Father Was Told," 285, 290.

31. Salome Ka&&ak Qalasiq, as cited in Kappianaq et al., *Dreams and Dream Interpretation*, 45. On Inuit names, see also Kublu, "Changing Perspectives"; Irniq, "Inuit Naming"; Alia, *Names and Nunavut*; Aupilaarjuk et al., *Cosmology and Shamanism*, 9–10. On names and home, see Trott, "Ilagiit and Tuqłuraqtuq," 12–17; Nuttall, *Arctic Homeland*, 68–69, 133–34; Søby, "Angerdlartoqut." The importance of dreams and one-year tradition is from Andrew Dialla, pers. comm.

32. Brody, *Other Side of Eden*.

33. Qitsualik, "Nunataaq."

34. Collignon, "Inuit Place Names and Sense of Place," in Stern and Stevenson, eds., *Critical Inuit Studies*, 204. On trails, see Aporta, "The Trail as Home."

35. See, e.g., Qikiqtani Inuit Association, *Community Histories: Pangnirtung*, 15; Boas, *Central Eskimo*, 54–62; Stevenson, *Cultural Persistence*, 74–75.

36. Larcom, *New England Girlhood*, 187–88; on Larcom, see also Matt, *Homesickness*, 50, 54–55. On the financial need to send children out to work, see, e.g., Bradbury, "Fragmented Family."

37. Melville, *Moby-Dick*, 37.

38. Hareven, "Home and the Family in Historical Perspective." On women's domestic labor, see Boydston, *Home and Work*. On one woman re-creating home in a new house, see Harris and Phillips, eds., *Letters from Windermere*, xvi. On the tension between staying put and moving on in agricultural societies, see Brody, *Other Side of Eden*.

39. Matt, *Homesickness*, chap. 4.

40. Dick, *Muskox Land*; Stuhl, *Unfreezing the Arctic*; Cruikshank, *Do Glaciers Listen?*; Cameron, *Far Off Metal River* (Cameron's concern is with Qallunaat understandings of the Arctic, but colonialism and Inuit history are central to her analysis); and Qikiqtani Inuit Association, *Achieving Saimaqatiqiingniq*, and the Qikiqtani Truth Commission's other reports available at qtcommission.ca. For the nineteenth and early twentieth centuries, see also Lowenstein, *Ultimate Americans*; Grant, *Arctic Justice*; Jensen et al., *Cultural Encounters at Cape Farewell*. Key works that deal with either Inuit or Qallunaat are too numerous to list here but see Works Cited.

41. On this topic, see, e.g., Cameron, *Far Off Metal River*, and the works of Julie Cruikshank, especially "Images of Society" and *Do Glaciers Listen?*

42. Some excellent starting points for learning more about Inuit knowledge from outside Inuit Nunangat (Inuit homelands) are the publications by Nunavut Arctic College, available through nacmedia.ca; the histories of the Qikiqtani region, 1950–75, produced by the Qikiqtani Truth Commission, available at qtcommission.ca; and the resources on Inuit Tapiriit Kanatami's website, itk.ca. Please see the methodological essay included in this volume for more on both my position as an outsider and issues surrounding Qallunaat research in Inuit Nunangat.

43. Environmental history is a vibrant and evolving discipline, but some of the foundational texts that originally sparked my interest are Worster, *Dust Bowl*; Cronon, *Nature's Metropolis*; and White, *Organic Machine*.

44. Much but not all this work falls into two categories: people affected by high-modern mega projects, and nineteenth-century settlers' "acclimatization societies" that reintroduced species from home. See, e.g., Parr, *Sensing Changes*; Valenčius, *Health of the Country*; Nash, *Inescapable Ecologies* and "The Changing Experience of Nature"; Loo, "Disturbing the Peace"; Carlson, *Home Is the Hunter*; Coates, *American Perceptions*; Dunlap, *Nature and the English Diaspora*, chap. 4; Ritvo, "Going Forth and Multiplying."

45. In the United States, see, e.g., Boydston, *Home and Work*; Jane E. Simonsen, *Making Home Work*; Ryan, *Cradle of the Middle Class*; Romero, *Home Fronts*; Edwards, *Gendered Strife*; Wilson, *Ye Heart of a Man*; Norling, *Captain Ahab Had a Wife*. For works that deal with Qallunaat settlers remaking and missing home in the nineteenth century, see, e.g., Johnson, *Roaring Camp*; Holland, *Home in the Howling Wilderness*; Valenčius, *Health of the Country*; Matt, *Homesickness*.

46. There is considerable work on time, modernity, and space/place, but most of it does not deal directly with ideas of home. See, e.g., Thompson, "Time, Work-Discipline, and Industrial Capitalism"; Glennie and Thrift, "Reworking E. P. Thompson" and "Time-Geography"; Lynch, *What Time Is This Place?*; Kern, *Culture of Time and Space*; Warf, *Time-Space Compression*; Harvey, "Between Space and Time"; Friedland and Boden, *NowHere*. More specifically about home,

see Hareven, *Family Time and Industrial Time*, and Matt, *Homesickness*, esp. 174. On the Arctic and time, see Stuhl, "Cold Places" and *Unfreezing the Arctic* (esp. the introduction).

47. On Inuit concepts of time, see Stern, "Upside Down"; MacDonald, *Arctic Sky*, 192–208; Nagy, "Time, Space, and Memory," in Stern and Stevenson, eds., *Critical Inuit Studies*.

48. Matt, *Homesickness*, 46, 55, 91–92.

49. Sonne, "The Acculturative Role"; Laugrand, *Mourir et renaître*, chap. 10.

50. Hatton and Harvey, *Newfoundland*, 170. For an argument that people deliberately stayed on the margins of states, see Scott, *Art of Not Being Governed*. On Inuit and the Victorian progress narrative, see also Spufford, *I May Be Some Time*, 205–22.

51. Thanks to Jackson Lears, Susan Schrepfer, Ann Fabian, Paul Clemens, and Chris Trott for discussions about the various concepts of time. On people's sense of place within time and history, see Walter Johnson, "Time and Revolution in African America." On "familiar, looping sorts of rupture," see Julie Livingston, in Eustace et al., *AHR Conversation: The Historical Study of Emotions*, 1520–21.

52. W. F. Campbell, as cited in Blake, ed., *Arctic Experiences*, 477.

53. My retelling of this story is based on the following accounts: Mary-Rousselière, *Qitdlarssuaq*; Uvdloriaq, "The Narrative of Qitdlarssuaq," NWT Archives, N-1998-047; Petersen, "The Last Eskimo Immigration"; Harper, "Qillarsuaq"; Rasmussen, *People of the Polar North*; McClintock, *Voyage of the Fox*, 121–22; Davis and Hall, *Narrative of the North Polar Expedition*, 450–51. See Schledermann, *Voices in Stone*, 148–65 for an archaeologist's account of excavating a site possibly used by Qillarsuaq's party.

54. They may have crossed Tallurutiup Imanga (Lancaster Sound) or Barrow Strait. Mary-Rousselière, *Qitdlarssuaq*, 38; Uvdloriaq, "The Narrative of Qitdlarssuaq," 14/2.

55. Mary-Rousselière, *Qitdlarssuaq*, 52; Knud Rasmussen, *People of the Polar North*, 28. On wood above the treeline, see Claire Alix, ed., "Arctic Peoples and Wood."

56. Merqusaq as cited in Rasmussen, *People of the Polar North*, 28.

57. Merqusaq as cited in Rasmussen, *People of the Polar North*, 29.

58. Uvdloriaq, "The Narrative of Qitdlarssuaq," 40/1.

59. Merqusaq as cited in Rasmussen, *People of the Polar North*, 33.

60. Merqusaq as cited in Petersen, "Eskimoernes sidste indvandring," 387. My translation from Danish.

61. Trott, "Ilagiit and Tuqłuraqtuq," 14, and pers. comm.

Chapter One

1. *Antelope* 1860–61, July 3–5, 1860, NBWM. Peter's recorded punishment was minor: he was sent aloft for two hours, then continued his regular activities in irons for a day, until the ship moved into open water. There were two foremast hands named "Peter" aboard: Peter Sullivan and Peter McBein. *Antelope* 1860–61 Crew List, NARA MA. On watches, see Creighton, *Rites*, 61.

2. On the wealth of "Arcticana" consumer goods, see Robinson, *Coldest Crucible*, 3; Potter, *Arctic Spectacles*, 4. On the British coverage of McClintock's findings in 1859–60, see Cavell, *Tracing the Connected Narrative*, 12 (and all of chap. 1). On Kane's popularity, see Robinson, *Coldest Crucible*, chap. 2; Potter, *Arctic Spectacles*, chap. 5.

3. *Antelope* 1860–61, July 3, 1860, NBWM.

4. For an overview of demographic estimates in Cumberland Sound, see Goldring, "Southeast Baffin Island," chap. 7.0.

5. Tyson, *Cruise*, 26. See also Faulkner, *Eighteen Months*, 88–92.

6. On whalers vs. explorers, see Ross, *Distant and Unsurveyed*, xviii, and "The Type and Number of Expeditions in the Franklin Search." On ships in view, see *Antelope* 1860–61, July 3, 1860, NBWM. Whaling was a huge industry. Dolin estimates the number of men on American whaling ships in the mid-1850s at 20,000, in *Leviathan*, 221. For maps of worldwide American whaling, see Smith et al., "Spatial and Seasonal Distribution of American Whaling."

7. Goldring, "Southeast Baffin Island," 7.20.

8. Meeka Mike mentioned eight recorded seasons in the South Baffin communities of Kimmirut and Cape Dorset (public lecture at the University of Calgary, November 5, 2010). July Papatsie lists eight seasons; von Finckenstein, ed., *Nuvisavik*, 21. The late South Baffin Elder Pauloosie Angmarlik named seven seasons; Laugrand and Oosten, eds., *Introduction*, 127–28. I spell the names of the seasons with the orthography in Spalding, *A Multi-Dialectical Outline Dictionary*.

9. Burt, *Barrenland Beauties*, 56–57.

10. On explorers overwintering, see Savours, *North West Passage*, 6; Cavell, "Going Native," 27.

11. Blake, ed., *Arctic Experiences*, 89; Barron, *Old Whaling Days*, 37 (overwintering location); Stevenson, *Cultural Persistence*, 75. On the lay system and wages, see Davis et al., *In Pursuit*, chap. 5. Qimiqsuut has also been written as Kemisuack, Kingmiksoo, Kingmiksok, Keimuksoke, Qimmiqsuut, and other variations. I use the spelling on Inuit Heritage Trust's draft map, which marks Qimiqsuut as an "old camp" on an island (Nimigen Island) near Uttuusivik; see ihti.ca/eng/place-names/pn-index.html. The Qimiqsuut of the early whaling period may have been located further north on the same island (Chris Trott, pers. comm.).

12. One was reportedly English and one from the Azores. Hall, "Journal," February 1860, item 4, Charles Francis Hall Collection, NMAH (this collection hereafter cited as CFHC NMAH). Thomas Evans may have been American, not English; he is listed as born in New York on the crew list in New London Crew Lists: 1803–78, MSM.

13. On crews, see Creighton, *Rites*, 28–31, 128; Davis et al., *In Pursuit*, 179–80.

14. Blake, ed., *Arctic Experiences*, 89; interview with George Tyson, in Hall, "Journal," February 1860, item 4, CFHC NMAH; Hall, *Arctic Researches*, 248 (slightly more details in interview). Blake says the roof was canvas; Hall says sealskin.

15. Pauloosie Veevee, interview, August 28, 2008, AVC; Hall, *Arctic Researches*, 461.

16. Budington remembered sixteen whales and Tyson seventeen. The baleen would have brought in approximately $8,120. Budington recorded that they took 16,000 pounds of whalebone, and whalebone sold for 50.75 cents per pound in New Bedford that year. (This supports reports of the whales being small, as Ross in "Annual Catch" (107) estimates an average of 1,392 pounds of bone per whale taken by Americans in the Davis Strait.) Budington does not state the amount of oil, but a conservative estimate of twenty tons from each of sixteen adult male bowheads, and 240 gallons of whale oil to a ton, makes 76,800 gallons of oil, which would have sold for 68 cents per gallon in New Bedford, bringing in $52,224. In the United States it was reported that only 60 tons of oil, from six whales captured before the *McLellan* was lost, were on board the *Truelove*, but Budington's and Barron's accounts contradict this. Regardless, the voyage was an incredible return on investment, given that the *McLellan* was likely insured. Colby, *For Oil*, 96; Blake, ed., *Arctic Experiences*, 90; Davis et al., *In Pursuit*, 374 (oil prices) and 377 (whalebone prices); Harper, *Pangnirtung*, 12, 14 (oil from male and female bowheads); Scoresby, *An Account of the Arctic Regions*, 2:403 (gallons of oil to a ton of blubber); *Whalemen's Shipping List*, November 30, 1852 (reported oil); Boydston, *Home and Work*, 134 (New York City living costs in 1860).

17. Blake, ed., *Arctic Experiences*, 89–90.

18. Barron, *Old Whaling Days*, 37–44.

NOTES TO PAGES 6–13

19. Barron, *Old Whaling Days*, 42–43. I'm not sure how the lay system would have played out in this unique situation. Busch says whalers whose ships were wrecked were "likely to receive nothing at all," in *Whaling Will Never Do For Me*, 9.

20. Interview with William Sterry, in Hall, "Journal," item 4, CFHC NMAH (quotes). Hall often mentions Sterry in *Arctic Researches*, e.g. xxiii, 32–33, 54–55, 252, 260. For Tyson's whaling voyages, see Blake, ed., *Arctic Experiences*, 77–99, and Tyson, *Cruise*. For Budington's whaling voyages, see Colby, *For Oil*, 95–103.

21. The two American ships in 1853–54 were the *Amaret* and *Georgiana*; see National Maritime Digital Library, "American Offshore Whaling Voyages."

22. On Captain William Penny, see, e.g., Cavell, *Tracing the Connected Narrative*, 171–72, 194–99. These sections also relate to the debates in the British press regarding how long Franklin's crew could live off the land.

23. Hall, "Journal," February 16, 1860, item 4, CFHC NMAH (emphasis in original).

24. See Hall's list of his Arctic books in Hall, "Journal, with Preparations for the First Expedition," item 2, CFHC NMAH.

25. Unless otherwise noted, all information about this desertion comes from: Hall, "Journal," August 7, 1860, item 13, CFHC NMAH; Hall, *Arctic Researches*, 90–97; *Ansel Gibbs* 1860–61, August 6–12 and October 11, 1860, NBWM; *Daniel Webster* 1860–63, August 5, 1860, NBWM; "A Harrowing Tale," *St. Johns Daily News*, December 6, 1861. The notebook containing John Sullivan's original testimony is item 17, CFHC NMAH. It is also reprinted in Hall, *Arctic Researches*, and Ross, *Arctic Whalers*.

26. Creighton, *Rites*, 94, 108–9, 144, 146.

27. Creighton, *Rites*, 61; Davis et al., *In Pursuit*, 285.

28. *Andrews* 1867, October 15 and 18 and November 2, 1867, NBWM. This was described as "freezing the boats up."

29. *Andrews* 1865–67, November 11 and 15, 1865, HL.

30. *Daniel Webster* 1860–63, November 25 and December 3, 1860, NBWM; *Antelope* 1860–61, November 23 and December 3, 1860, NBWM; *Andrews* 1865–67, November 1, 1865, HL.

31. Stevenson, *Cultural Persistence*, 81.

32. Unless otherwise noted, all information about the *Andrews* shipwreck comes from: *Andrews* 1867, November 15–28, 1867, NBWM; *Andrews* 1865–67, November 14–December 8, 1867 and the summary entry after February 2, 1868, HL.

33. The crew list for this voyage has not survived, but on previous voyages the *Andrews* had twenty-five to thirty men on board. See crew lists for 1863, 1865, 1866, and 1867 (summer voyage) at NARA MA.

34. *Milwood*, 1867–68, March 4, 1868, NBWM.

35. *Milwood*, April 1868, NBWM.

36. See *Ansel Gibbs* 1860–61, October 9–11, 1861, NBWM; *Antelope* 1865–66, October 8, 1866, NBWM; *Milwood* 1867–68, January and April 1868, NBWM.

37. For a whaler's depiction of the Qallunaat community helping during a ship fire in 1866, see Cardno, *The Whaler Dublin in Flames*, in Cardno, *A Whaler's Tale*, 13, or online at http://www.nefa.net/archive/peopleandlife/sea/tale.htm. Compare this activity at Naujaaqtalik with the desolation depicted in fig. 1.3.

38. Kowjakuluk, interviewed in Dorothy Eber, "Inuit Memories of the Whaling Days," Ms. IV-C-138M, CMH, 48; Ross, *Distant and Unsurveyed*, 102.

39. Bockstoce, *Whales, Ice, and Men*, 271–72.

40. *Andrews* 1865–67, October 17, 1867, HL.

41. Tyson, *Cruise*, 60–61, 123.

42. Creighton, *Rites*, 30; Davis et al., *In Pursuit*, 47. Stifling quote is L. E. Borden, as cited in Ross, "Canadian Sovereignty," 97–98.

43. George A. Comer, in interview with Fred Calabretta, *MSM*.

44. Chris Trott, pers. comm.

45. On soap and washing, see Boas in Müller-Wille, ed., *Franz Boas*, 61.

46. See, e.g., Spufford, *I May Be Some Time*, 199–200. Franz Boas alluded to this pervasive stereotype when he wrote of his first visit to Inuit homes in Cumberland Sound, "Not as dirty as I thought." Boas, quoted in Müller-Wille, ed., *Franz Boas*, 73.

47. *UD* 1864–69, undated entry, PPL.

48. See, e.g., *Andrews* 1865–67, November 30, December 1 and 29, 1865, HL; *Daniel Webster*, December 8, 1860, September 5, 1861, NBWM; Faulkner, *Eighteen Months*, 251.

49. Sullivan in Hall, *Arctic Researches*, 94.

50. On clothing, see Ross, *Distant and Unsurveyed*, 104–8. Faulkner bought a complete winter outfit for five bottles of molasses and six plugs of tobacco; *Eighteen Months*, 92. Ambrose Bates also mentions whalers wearing skin clothing in a poem in *UD* 1864–69, February 28, 1868, PPL.

51. Boas in Müller-Wille, ed., *Franz Boas*, 212.

52. Tyson, *Cruise*, 36.

53. *Isabella* 1867–68, October 13, 1867, NBWM; Goldring, "Whaling-Era Toponymy," 32.

54. *Ansel Gibbs* 1860–61, September 11, 1860 and January 27, 1861, NBWM; *Andrews* 1867, September 30, 1867, NBWM.

55. Ross, *Arctic Whalers*, 159.

56. Ross, *Arctic Whalers*, 173.

57. Tyson to "My Dear Wife," August 20, 1877, in George E. Tyson Papers, NARA MD.

58. Jean Bullard, in interview with Fred Calabretta, speaking about Captain George Comer, *MSL*.

59. *Milwood* 1867–68, February 1868 (quotes), August 14, 1868 (man who stayed), NBWM.

60. Adam Bek, "Akilinermut . . ." 8–9, 11 [page numbers refer to the partial draft translation courtesy of Kenn Harper]. For other disputes involving Inuit women, see Cardno, *Whaler's Tale*, 25; Faulkner, *Eighteen Months*, 203–8.

61. Hantzsch, *My Life*, 218–19.

62. *Milwood* 1867–68, June 1867 and June 1868, NBWM.

63. *Milwood* 1867–68, April 2, 1867, NBWM; Tyson, *Cruise*, 58. Margaret Penny, the wife of a Scottish whaling captain, bestowed more positive descriptions than most men, including "terrifick grand" and "a pleasant valley" and "a fine beach for fishing." Ross, *Distant and Unsurveyed*, 19, 34.

64. The average of two deaths does not account for many men who were shipped home on other vessels because they were seriously ill; presumably not all of them recovered. Nor does it include John Sullivan and the other deserters. A broader study of the American whaling industry did not contain a single recorded voyage without at least one death. Creighton, *Rites*, 198.

65. *Isabella* 1867–68, June 12, 1868, NBWM.

66. For all deaths in surviving American logbooks and journals from Cumberland Sound in the 1860s, see: *Milwood* 1867–68, June 1868, NBWM (one drowning, one died after eating whale skin "after being long without food"); *Antelope* 1865–66, May 4, 1865, NBWM (drowning); *Andrews* 1865–67, June 29, 1866 (consumption), July 24, 1866 (scurvy), HL; *Andrews* 1867, July 2, 1867 (unknown disease, does not seem to be contagious), NBWM; *Black Eagle* 1860–61, September 22, 1860 (listed as dropsy [edema] on crew list return), August 1, 1861 (likely a

NOTES TO PAGES 17–20

contagious disease, listed as kidney complaint in crew list return), September 1, 1861 (disease, listed as kidney complaint in crew list return), NBWM; *Daniel Webster* 1860–63, April 24, 1861, March 20, 1862, July 19, 1862, August 9, 1862 (all scurvy), NBWM; *Ansel Gibbs* 1860–61, April 18, 1860 (drowning), April 17, 1861 (mentioning a death from an illness on the *Hannibal*, for which no logbook survives), October 29, 1861 (mentioning a death from consumption on the *Hannibal*), NBWM; *Antelope* 1860–61, September 18, 1861 (drowning), NBWM; *Isabella* 1867–68, May 24, 1868 (drowning), NBWM. I know of three recorded freezing deaths of Qallunaat whalers on Baffin Island, just not of Americans in Cumberland Sound. See Cardno, *Whaler's Tale*, 18–19, 32 (two Scottish whalers and an Inuit companion); Hall, *Arctic Researches*, 225–42 (whaler from an American ship south of Cumberland Sound).

67. Tyson, *Cruise*, 57.

68. *Milwood* 1867–68, December 1867 (poem), July 4, 1868 (list of voyages), NBWM. His other recorded voyages were in 1851, 1863, 1865, 1867, and 1868; see New Bedford Whaling Museum, "Whaling Crew List Database." For US employment see Ancestry.com, *US IRS Tax Assessment Lists, 1862–1918*, State of Connecticut Third Collection District, December 1864; Ancestry.com, *1880 United States Federal Census*, Sterling, Windham, Connecticut, Enumeration District 147. On farming roots, see Ancestry.com, *1850 United States Federal Census*, Foster, Providence, Rhode Island. For his death, see "Ambrose H. Bates," *Find a Grave*, http://forums.findagrave.com/cgi-bin/fg.cgi?page=gr&GScid=2233379&GRid=84966236&.

69. *Milwood* 1867–68, August 1867, NBWM.

70. Bates's description in *UD* 1868–69, June 1865, PPL; Creighton, *Rites*, 29, 173.

71. *Milwood* 1867–68, September 11, 1868, NBWM.

72. *Milwood* 1867–68, January 1868, NBWM. See also Cull, "A Description of Three Esquimaux," 217–18. Consider the Inuk Hans Hendrik's converse assessment of the United States' dependence on agriculture: "How wonderful that all these people subsist from the trifle that the soil produces." Hendrik, *Memoirs*, 76.

73. Spufford, *I May Be Some Time*, 194.

74. David M. Henkin, *Postal Age*.

75. *Milwood* 1867–68, December 25, 1867 (letters), June 1868 (ice field), NBWM.

76. Jacopoosie Peter, public lecture at the University of Calgary, November 5, 2010 (means to hear); Kappianaq and Nutaraq, *Travelling and Surviving*, 109, 182 (quote about ice forming); Taamusi Qumaq, as cited in Bordin, "La nuit inuit," 51 (dictionary; translation from French is mine).

77. *Milwood* 1867–68, April 1868, May 1867, NBWM.

78. Ross, *Arctic Whalers*, 183.

79. *UD* 1868–69, "What Do We Love" [undated poem by Bates], PPL.

80. Mark Nuttall coined the term "memoryscape" in regards to the area around Kangersuatsiaq in Greenland. Nuttall, *Arctic Homeland*, chap. 4.

81. See Collignon, *Knowing Places*, 90.

82. Ludger Müller-Wille and Linna Weber Müller-Wille, "Inuit Geographical Knowledge One Hundred Years Apart," in Stern and Stevenson, eds., *Critical Inuit Studies*, 226–27. Maps from the Inuit Heritage Trust Place Names Program are available at ihti.ca. On place names, see also Kappianaq and Nutaraq, *Travelling and Surviving*, 141; Collignon, "Inuit Place Names and Sense of Place," in Stern and Stevenson, eds., *Critical Inuit Studies*, 187–205.

83. Goldring, "Whaling-Era Toponymy."

84. Laugrand et al., eds., *Apostle to the Inuit*, 397–468.

85. *Andrews* 1865–67, December 24, 1865, HL.

86. Tyson, *Cruise*, 57.

87. See, e.g., *Ansel Gibbs* 1860–61, December 2, 26, and 28, 1860; January 23, 1861; February 1, 20, 22, and 28, 1861, NBWM; *Antelope* 1860–61, January 22 and February 22, 1861 (theater), NBWM; *Andrews* 1865–67, December 16, 1865, March 23, 1868, HL; Faulkner, *Eighteen Months*, 187, 230–36. For pastimes on overwintering Arctic expeditions, see, e.g., Savours, *North West Passage*.

88. *Milwood* 1867–68, December 25, 1867, NBWM. On Inuit Christmas celebrations in a later period, see Laugrand and Oosten, "*Quviasukvik*."

89. Faulkner, *Eighteen Months*, 230 (seals), 236 (tobogganing). On seal at Christmas, see also *Andrews* 1865–67, December 25, 1867, HL. On Inuit learning English, see Boas in Müller-Wille, ed., *Franz Boas*, 91.

90. Tyson, *Cruise*, 103–4. In 2006, Pangnirtung Elder Inuusiq Nashalik also named activities in *upingaksaaq* as his favorite time of year (pers. comm.).

91. Pielou, *Naturalist's Guide*, 68.

92. *Ansel Gibbs* 1860–61, May 9, 11, 1861, NBWM.

93. Sullivan in Hall, *Arctic Researches*, 94.

94. For provisions lists, see *Daniel Webster* 1860–63, NBWM, and *Isabella* 1867–68, NBWM. On whaling diets, see Creighton, *Rites*, 126–27; Davis et al., *In Pursuit*, 48. On vitamin C in Inuit foods, see Geraci and Smith, "Vitamin C," esp. 137. On vitamin C content of today's Qallunaat foods, see United States Department of Agriculture, *USDA Food Composition Database*, https://ndb.nal.usda.gov/.

95. *A. R. Tucker*, 1891–92, June 28, 1891, PPL. This was in Hudson Bay, another whaling destination in the Canadian Eastern Arctic.

96. On scurvy, see Roger K. French, "Scurvy," in Kiple, ed., *Cambridge Historical Dictionary of Disease*, 295–97; Feeney, *Polar Journeys*, 8; Ross, *Distant and Unsurveyed*, 144–62; Bown, *Scurvy*.

97. Bown, *Scurvy*, 33–34.

98. *Daniel Webster* 1860–63, February 14, 23, 1861; most of April 1861; May 16, 1861; NBWM.

99. *Daniel Webster* 1860–63, December 10, 1860, April 30, 1861, NBWM.

100. *Black Eagle* 1860–61, February–May 1861 (seals), June 8 and August 19, 1861 (caribou), NBWM.

101. Colby, *For Oil*, 136–37. Ross details Scottish captain William Penny's extensive efforts to avoid scurvy in *Distant and Unsurveyed*, 157–62.

102. *Daniel Webster* 1860–63, April 24, 1861, NBWM; *Pioneer* (no log but see *Ansel Gibbs* 1860–61, April 23, 1861, NBWM); *Hannibal* (likely scurvy, see *Ansel Gibbs* 1860–61, April 17, 1861, NBWM); Ross, *Distant and Unsurveyed*,162. The *Black Eagle* only overwintered for one year so its crew was in less danger, but there were fewer ships the second year so it would likely have been possible for the *Daniel Webster*'s captain to procure fresh meat if he had decided to make it a priority.

103. Ross, *Distant and Unsurveyed*, 146–48, 152–55.

104. Robert Scott's notes from Edward Atkinson's lecture, quoted in Carpenter, *History of Scurvy*, 153.

105. *Milwood* 1867–68, April 1868, NBWM.

106. *Ansel Gibbs* 1860–61, January 28–30, 1861, February 9, 1861 (sleighing quote), March 7, 1861, May 16, 1861 (earth cure), May 19, 1861 (deer and seals), May 23, 1861 (doing better), NBWM.

107. *Ansel Gibbs* 1860–61, January 31 and February 2, 1861, NBWM.

108. Ross, *Distant and Unsurveyed*, 152.

109. On pigs on Cumberland Sound whaling ships see, e.g., *Andrews* 1865–67, November 7,

1865, November 22, 1865, HL; *Daniel Webster* 1860–63, November 9, 1860, NBWM. See also Ross, *Distant and Unsurveyed*, 150–51.

110. See, e.g., *Andrews*, 1865–67, July 13, 1867, HL.

111. References to char appear in nearly all the logbooks from late July through September. See, e.g., *Daniel Webster* 1860–63, August 1 and 10, 1860, August 20, 1861, NBWM; *Ansel Gibbs* 1860–61, July 31–August 3, 1860, NBWM. The *Ansel Gibbs* records clamming from August to November of 1860 (e.g., *Ansel Gibbs* 1860–61, August 20, 23, 1860; September 18, 1860; November 14, 1860, NBWM). For shootings of small game, see, e.g., *Ansel Gibbs* 1860–61, March 8, 1861, January 10, 1861, October 9, 1861, June 11, 1861, NBWM; *Andrews* 1867, August 26, 1867, NBWM. On attempts of whalemen to catch seals, see *Black Eagle* 1860–61, December 12, 1861, NBWM; Faulkner, *Eighteen Months*, 214–22, 253–54. For berries, see Ross, *Distant and Unsurveyed*, 39–40. For eggs and ducks, see *Andrews* 1865–67, July 6, 1866, HL.

112. On length of bowheads, see Ross, *Distant and Unsurveyed*, 84–85; Colby, *For Oil*, 146.

113. Ross, *Arctic Whalers*, 85–86; Ross, *Distant and Unsurveyed*, 157. Both whale skin and blubber contain vitamin C, but the skin is an especially strong source. Geraci and Smith, "Vitamin C," 137.

114. Tyson, *Cruise*, 70–71, 89, 93.

115. Sterry was recounting the voyage of the *Georgiana* in 1855. See Colby, *For Oil*, 97; Hall, "Journal," item 4, CFHC NMAH. For more on not assuming feelings are mutual, see Goldring, "Historians and Inuit," 520–21.

116. Hall, "Journal," June 30, 1860, item 12, CFHC NMAH; Hall, *Arctic Researches*, 39, 225–42; *George Henry* 1860–61, March 17, 1861, NBWM. Ross notes some whalers refusing antiscorbutics in Hudson Bay, in *Distant and Unsurveyed*, 156–57.

117. Boas in Müller-Wille, ed., *Franz Boas*, 197. On scurvy in the US, see, e.g., Ellis, "Presence and Absence."

118. Stevenson, *Cultural Persistence*, 74–76, 83–85; Goldring, "Southeast Baffin Island," 7.3; Goldring, "Inuit Economic Reponses," 159 (food shortages); Boas in Müller-Wille, ed., *Franz Boas*, 128–31. On contact and disease in what is now Canada, see, e.g., Hackett, *Very Remarkable Sickness*; Daschuk, *Clearing the Plains*; Piper and Sandlos, "A Broken Frontier"; Langston, "Thinking like a Microbe."

119. Dolin, *Leviathan*, 271. Most studies comparing wages for whalers and shore laborers do not appear to factor in the room and board provided on whaling ships. In the 1860s, agricultural laborers were charged between $96 and $180 per year for food and lodgings. Room and board for a single frugal adult male in 1860 in New York City was $208. Davis et al., *In Pursuit*, 180–85; Wright, *Comparative Wages*, 47 (agricultural laborers); Boydston, *Home and Work*, 134 (New York). For a sample of the earnings crew members took home and what they purchased on credit, see Accounts of the *Black Eagle*, NBWM.

120. Rugh and Shelden, "Bowhead Whale," 131; Davis et al., *In Pursuit*, 22, 24.

121. Ross, *Arctic Whalers*, 163.

122. On recent changes to the floe edge, see Laidler, "Ice through Inuit Eyes," 312–13.

123. Stevenson, *Cultural Persistence*, 80. On Tessuin, see also Ross, *Distant and Unsurveyed*, 188; Goldring, "Inuit Economic Responses," 157 (refers in part to Tyson's account in *Cruise*).

124. Davis et al., *In Pursuit*, 25.

125. Blake, ed., *Arctic Experiences*, 87.

126. A similar situation occurred in Faulkner, *Eighteen Months*, 289–90. On sledding and towing whales see Ross, *Distant and Unsurveyed*, 83; Ross, *Arctic Whalers*, 163.

127. Blake, ed., *Arctic Experiences*, 87. On processing whalebone, see Creighton, *Rites*, 72.

128. Unidentified source, quoted in Davis et al., *In Pursuit*, 17n19.

129. Faulkner, *Eighteen Months*, 87.

130. Davis et al., *In Pursuit*, chap. 9, esp. 342–51, 368.

131. Blake, ed. *Arctic Experiences*, 90; Agee Temela in Eber, "Inuit Memories of the Whaling Days," CMH, 73; *Andrews* 1865–67, October 7, 1865 (meatballs), HL; Faulkner, *Eighteen Months*, 172 (whale steak).

132. Most whaling vessels were 100–150 feet in length, and the Arctic vessels tended to be barks on the smaller end of the range. Creighton, *Rites*, 28.

133. Scoresby, *Account of the Arctic Regions*, 2:455; Blake, ed., *Arctic Experiences*, 85 (boat in mouth).

134. Kowjakuluk, interviewed in Eber, "Inuit Memories of the Whaling Days," CMH, 42.

135. Faulkner, *Eighteen Months*, 295.

136. Creighton, *Rites*, 69.

137. *Andrews* 1865–67, June 5, 1866, HL; Blake, ed., *Arctic Experiences*, 86.

138. *Milwood* 1867–68, February 1868, NBWM.

139. Blake, ed., *Arctic Experiences*, 89–90. See also Colby, *For Oil*, 86; Outerbridge, "My Voyage to Cumberland Inlet," 67–68.

140. *Milwood* 1867–68, April 1868, NBWM.

141. *Milwood* 1867–68, September 10, 1868.

142. Kumlien, *Contributions*, 64–65 (describes decline of fishery); Stevenson, *Cultural Persistence*, 81–83; Goldring, "Inuit Economic Responses," 153.

143. See, e.g., Rosenow, *Death and Dying*. On working-class diets in New York, see Burrows and Wallace, *Gotham*, 477.

144. The small existing literature on ships as homes focuses on women—especially female tourists—re-creating their domestic life at sea. Pagh, *At Home Afloat*; Ryan, "Our Home on the Ocean." Most uses of "home" in whaling records are reflexive and are used either to refer to the ship or to the United States, e.g., "sent Charles Smith home [to the United States] in the Brig Georgiana of New London he having some of his toes froze"; or, the men were outside playing games with other crews and "all came home [to this ship] but three." *Daniel Webster* 1860–63, September 5, 1861, NBWM; *Ansel Gibbs* 1860–61, February 23, 1861.

145. Hasty and Peters, "The Ship in Geography," 664–65.

146. Blake, ed., *Arctic Experiences*, 77, 91–92; Tyson, Diaries, [undated, appears to be a draft of the manuscript for *Arctic Experiences*], box 2, George E. Tyson Papers, NARA MD.

147. Blake, ed., *Arctic Experiences*, 375.

148. Tyson, *Cruise*, 57.

149. "Captain Budington: His Home in Connecticut," *New York Times*, August 4, 1873; Eber, *When the Whalers*, 70; Decker, *Whaling City*, 83; Sally Motycka, "About This Artifact: Cross-Cultural Clothing," on Mystic Seaport website: http://educators.mysticseaport.org/artifacts/spicer_inuit_outfit/; "Baleen cane with antler grip," object 1956.921 in Mystic Seaport collections, Bullard interview with Fred Calabretta, *MSL*. For Spicer's voyages, see National Maritime Digital Library, "American Offshore Whaling Voyages." This lists only the voyages where he sailed as captain, so he was likely visiting the Arctic earlier than 1863.

150. Thanks to Meg Stanley for conversations about home as memory and mobile networks.

151. Harper, "William Duval"; Peteroosie Sivutiksaq Karpik, interview with the author, September 8, 2008, AVC (he carries the name).

Chapter Two

1. I refer to this Inuit couple as Ipiirvik and Hannah. I wanted to use their Inuktitut names, but the correct version of Hannah's name is not known today. It has been rendered as Tookoolito, Tackritow, Taqulittuq, and many other variations. Her husband either chose or was given the English name Joe. His Inuktitut name, Ipiirvik, was mistakenly spelled Ebierbing, Eberbing, and Epiopee. Another of his names may have been corrupted as Hackboch, Harboch, or Harkbah—these were the names recorded in England. The couple likely had other Inuktitut names they never shared with outsiders, and they probably did not address each other by name. Instead they would have used kinship terms or other words reflecting their relationship—thereby expressing their identities as part of a larger community, rather than as fixed and isolated individuals. Kenn Harper, pers. comm.; Ross, *Distant and Unsurveyed*, 57 (names in England); Trott, "Ilagiit and Tuqłuraqtuq," 14–15 (multiple names). Hannah and Ipiirvik were both probably born in the late 1830s. See "The New Arctic Discoveries," *New York Herald*, September 15, 1862; Nourse, ed., *Narrative*, 443; Ross, *Distant and Unsurveyed*, 54.

2. Hall, *Arctic Researches*, 562. Ipiirvik and Hannah were at Cape Farrington, in the Tuapajjuaraarjuit area according to the map of the Iqaluit area at ihti.ca.

3. Thanks to Kenn Harper for sharing his decades of collected information on Inuit travelers. For more on Inuit abroad, see, e.g., Eber, *When the Whalers*, chap. 4; Greenblatt, *Marvelous Possessions*, chap. 4; Harper, *Give Me My Father's Body* and *In Those Days*; Harbsmeier, "Bodies and Voices"; Potter, *Arctic Spectacles*, chap. 8; Ross, *Distant and Unsurveyed*; Rowley, "Eenoolooapik"; Thrush, "The Iceberg and the Cathedral" and *Indigenous London*; Vaughan, *Transatlantic Encounters*.

4. Hall, *Arctic Researches*, 562. Relatives included Hannah's brother Inuluapik, and Ipiirvik's half-brother Italoo Enoch and cousin Kud-lup-pa-mune or Abbott.

5. Colby, *For Oil*, 96; Barron, *Old Whaling Days*, 42 (Hannah teaching him some Inuktitut in 1852); Harper, "Ebierbing, Hannah [Tookoolito] and Joe," in Nuttall, *Encyclopedia of the Arctic*, 520; Nourse, ed., *Narrative*, 443. For more on Hannah and Ipiirvik see, e.g., Potter, *Arctic Spectacles*, chap. 8; David, *Arctic in the British Imagination*, 135–37; Robinson, *Coldest Crucible*, 73–74; Harper, *In Those Days*, 61–82; Ross, *Distant and Unsurveyed*, esp. 48–58.

6. We can't be sure of Hannah and Ipiirvik's motivations for going abroad in 1862 or even if they wanted to go; we only know what Qallunaat recorded. One record says that Hannah wanted to prevent Ipiirvik taking a second wife; see Nourse, ed., *Narrative*, 444. The whaler Ambrose Bates later told a story of a feud involving an Inuk named "Tookalooky" in his journal from the *Milwood*, 1867–68, January 15, 1868, NBWM. The story does not fit chronologically with Hannah's departure, however, and Bates may well have fabricated it.

7. Hannah to Sarah Budington, August 17, 1863 and April 1, 1864, Ms Eb47 M1159 and 1162, ICRC. On Sundays, see McCrossen, *Holy Day*.

8. MacDonald, *Arctic Sky*, 203–8.

9. J. J. Copp, "Handwritten Article about Sylvia Grinnell Little Punny," Ms Eb47 M1187, ICRC.

10. Barnum's American Museum was open "from sunrise until 10 p.m., seven days a week." Saxon, *P. T. Barnum*, 108.

11. William Guay to Charles Francis Hall, November 18, 1862, item 45, CFHC NMAH (cotton banner); "Advertisement for P. T. Barnum's Museum," item 47, CFHC NMAH; classified advertisement, *New York Times*, November 17, 1862. For more on Barnum's aggressive advertising strategies, see Henkin, *City Reading*, 82–83.

12. Classified advertisements, *New York Times*, November 9–22, 1862; Cook, *Arts of Deception*, 24, 139–40; Dennett, *Weird and Wonderful*, 35; Ryan, *Forgotten Aquariums of Boston*, 41–42 (belugas); Barnum, *An Illustrated Catalogue*; Adams, *E Pluribus Barnum*, 78; Martin, *White African American Body*, 101–3. African Americans were not regularly allowed in Barnum's museum until after the Civil War.

13. "The Esquimaux," *New York Times*, November 21, 1862. The podiums are noted in 1865 in "Disastrous Fire," *New York Times*, July 14, 1865.

14. "Advertisement for P. T. Barnum's Museum," item 47, CFHC NMAH.

15. Bogdan, *Freak Show*, 33.

16. On immigration statistics, see Burrows and Wallace, *Gotham*, 736.

17. Robinson, *Coldest Crucible*, 47, 69–70.

18. See, e.g., "The Esquimaux," *Daily Evening Bulletin* (San Francisco, CA), July 18, 1873; "When an Esquimau gentleman eats half a dozen tallow candles for his lunch and washes it down with a pint of lamp oil, can he be said to have made a light repast?" *San Francisco Daily Evening Bulletin*, June 4, 1875; "Many visitors; visit Peary's ship and see the Esquimaux," *Boston Daily Advertiser* September 28, 1897; "Our Interesting Esquimau Visitors," *Denver Evening Post* February 12, 1899. These stereotypes became more entrenched as the nineteenth century went on. On perceptions of Arctic Indigenous peoples see, e.g., Fienup-Riordan, *Eskimo Essays*; Trott, "The Dialectics of 'Us' and 'Other.'"

19. "Scrapbook of Newspaper Clippings, 1858–1863," item 3, CFHC NMAH; "Arctic Explorations—A Lecture by Captain C. F. Hall," *Daily Cleveland Herald*, October 22, 1869.

20. "Interesting and Genuine," *Milwaukee Daily Sentinel*, November 18, 1869.

21. Hall to Ira Hart, November 22, 1862, item 45, CFHC NMAH; Ross, *Distant and Unsurveyed*, 55–56 (England).

22. On clothing and status, see Cook, *Arts of Deception*, 160–61.

23. "Kad-lu-nah," *New York Times*, November 7, 1862. On Inuit and clothing, see also Spufford, *I May Be Some Time*, 232.

24. Hall to Sidney Budington, November 15, 1862, item 40, CFHC NMAH.

25. "Personal—Important Arrival," *New York Times*, November 11, 1862; "The Esquimaux—These Cheerful Strangers," *New York Times*, November 21, 1862.

26. Hall, *Arctic Researches*, 157–58, 161. On her English skills, see also Cull, "A Description of Three Esquimaux," 219.

27. Hall, "Journal," April 22, 1863, item 55, CFHC NMAH; Hall, *Arctic Researches*, 158, 161.

28. "Personal—Important Arrival," *New York Times*, November 11, 1862.

29. Maddox, *Removals*, 23–24.

30. See, e.g., Barnum's attempts to shape the "What Is It" performers, in Cook, "Of Men," 142–47.

31. "Personal—Important Arrival," *New York Times*, 11 Nov 1862; "Amusements," *New York Times*, 17 Nov 1862.

32. "The Esquimaux," *New York Times*, November 21, 1862.

33. On Inummariit, see Brody, *People's Land*; Stairs, "Self Image"; Wenzel, *Animal Rights*, 139–40. I am uncertain whether the term *inummariit* existed in the nineteenth century; today it is often used to contrast town Inuit with those who grew up on the land.

34. Ipellie, *Arctic Dreams and Nightmares*, vii.

35. Hall to Sidney Budington, November 26, 1862, and Hall to Sarah Budington, December 7, 1862, item 40, CFHC NMAH.

36. This aquarium was not Barnum's more famous one, but rather the shortlived inspiration

NOTES TO PAGES 45–51

of James Ambrose Cutting, an inventor who had already made and lost several moderate fortunes. Cutting had invented a new kind of beehive, and then, more famously, the ambrotype—a popular photographic technique of the period. He first became interested in ocean life when cruising around on the yacht he had purchased with his ambrotype patent earnings. "Death of an Inventor in an Insane Asylum," *New York Times*, August 14, 1867.

37. Classified Advertisement, *Boston Post*, December 1, 1862. Reprinted in Ryan, *Forgotten Aquariums of Boston*, 53. See also Classified Advertisement, *Boston Daily Advertiser*, November 29, 1862; Caladon Daboll to Hall, December 18, 1862, item 45, CFHC NMAH.

38. Sarah Gooll Putnam, "Diary 2," February 23, 1862, Massachusetts Historical Society, http://www.masshist.org/database/382.

39. Angmarlik in Nakasuk et al., *Interviewing*, 122. Also personal communication with residents of Pangnirtung.

40. Robinson, *Coldest Crucible*, 38–43 (Kane), 68–74 (Hall).

41. "Scrapbook of Newspaper Clippings," item 3, CFHC NMAH.

42. Chris Trott has made this argument about turn-of-the-century missionary portrayals of Inuit, in which the adoption of the nuclear family model and monogamy were seen as key evidence of conversion. Trott, "The Dialectics of 'Us' and 'Other,'" 186–87. See also Trott, "Ilagiit and Tuqłuraqtuq."

43. All quotes here from Hall, "Diary," February 16, 1860, item 1, CFHC NMAH.

44. M'Donald, *Narrative*, 21.

45. J. E. Nourse to J. J. Copp, December 14, 1876, Ms Eb47 M1183, ICRC. Hall to Henry Grinnell, June 8, 1870, Coll. 8, MSM.

46. "The Esquimaux," *Daily Evening Bulletin* (San Francisco, CA), July 18, 1873.

47. Hall to Sidney Budington, October 22, 1862, item 40, CFHC NMAH; Ross, *Distant and Unsurveyed*, 138; Hall, *Arctic Researches*, 157–58.

48. Hall to Sidney Budington, October 29, 1862, item 40, CFHC NMAH.

49. M'Donald, *Narrative*, 25; Cull, "A Description of Three Esquimaux," 215.

50. On Inuit disapproval of expressions of anger or hostility, see Briggs, *Never in Anger*, 328–37; Brody, *Other Side of Eden*, 42–44. See also Peter Kulchyski, "six gestures," in Stern and Stevenson, eds., *Critical Inuit Studies*, 159–60 (re: how "yes" can mean "no.")

51. "Scrapbook of Newspaper Clippings, 1858–1863," item 3, CFHC NMAH.

52. Cull, "A Description of Three Esquimaux," 216, 218–19; Hall, *Arctic Researches*, 157–59.

53. Hall to Sidney and Sarah Budington, April 22, 1863, item 40, CFHC NMAH.

54. Hall, "Journal," April 13, 1863, item 55, CFHC NMAH.

55. "The Esquimaux," *Daily Evening Bulletin* (San Francisco, CA), July 18, 1873.

56. Henry Baldwin Jr. to Hall, February 20, 1864, item 52; J. W. Allen to Hall, January 22, 1863, item 52; Louis Frobisher to Hall, August 17, 1863, item 52; Hall, "Journal," April 21, 1863, item 58, all from CFHC NMAH.

57. Hendrik, *Memoirs*, 78.

58. "Esquimau Joe: The Man Who Loved Captain Hall and Who Saved Captain Tyson's Party," *Daily Inter Ocean* (Chicago, IL), January 1, 1876.

59. Hall, "Journal," February 28, 1863 [entry date, although he must have added to it later], item 55, CFHC NMAH.

60. R. W. Seager to Hall, March 18, 1863, item 45, CFHC NMAH; Hall, draft of a letter sent to R. W. Seager, "Journal," February 12, 1863, item 55, CFHC NMAH.

61. Hall, "Notes for Lectures," item 46, CFHC NMAH.

62. Ittinuaq as cited in Aupilaarjuk et al., *Inuit Qaujimajatuqangit*, 182.

63. Ootoova et al., *Perspectives on Traditional Health*, 140 (cooling off).

64. Hall, "Speech given by Hall . . . 1869," item 104, CFHC NMAH. See also Berman, *All That Is Solid*; Schivelbusch, *Railway Journey*.

65. Note the word "nomad" passed into common English usage in the nineteenth century: "nomad, n. and adj.," *OED Online*, Oxford University Press, March 2017. On agricultural societies as the true nomads, see Brody, *Other Side of Eden*.

66. Hart to Hall, 11 November 11, 1862, item 45, CFHC NMAH; On Hart, see Twain et al., *Mark Twain's Letters*, 267n2.

67. "Photo for February 2004—The Rathbun Hotel," http://www.shs58.0rg/mike_paul/rathbun.html.

68. Burrows and Wallace, *Gotham*, 747.

69. Note added to letter from Hall to Sarah Budington, January 29, 1863, item 40, CFHC NMAH.

70. Burrows and Wallace, *Gotham*, 484, 827, 881; Anbinder, *Five Points*, 208, 213, 220, 235.

71. Burrows and Wallace, *Gotham*, 786–787.

72. Hall to Henry Grinnell, January 23, 1863, Coll. 8, MSM.

73. Hall to Sarah Budington, January 29, 1863; Hall to Sidney Budington, January 27, 1863, both in item 40, CFHC NMAH.

74. Hall to "Mr. Farie or Peale," February 15, 1863, item 52, CFHC NMAH; Hall to Sidney Budington, February 27, 18[63] [letter mistakenly dated 1862], item 40, CFHC NMAH; Hall to Grinnell, March 4, 1863, item 40, CFHC NMAH; Hall, *Arctic Researches*, 569; Hall "Journal," February 28 and March 2, 1863, item 55, CFHC NMAH. *Kipiniaqtuq* is from Ekho and Ottokie, *Childrearing Practices*, 133; see also Uqsuralik Ottokie, as cited on 86. On words for sickness, see Ootoova et al., *Perspectives on Traditional Health*, 77, 314, 195–96. *Qanimajuq* was used on South Baffin.

75. Christopher Trott, pers. comm.; Ekho and Ottokie, *Childrearing Practices*, 11, 35, 86. In 2000, Felix Pisuk of Rankin Inlet (Kangiq&iniq) spoke of *tarralikitat* as a type of *pisuqsauti*, "something that enabled you to do something": "*Tarralikitat*, butterflies, were also used as a means of helping you to come home. Someone would wipe the butterfly on the back of the parka of a young girl or boy." I do not know if Hannah and Ipiirvik would have shared this knowledge or seen it as having any relation to their son's name. Pisuk, as cited in Aupilaarjuk et al., *Inuit Qaujimajatuqangit*, 180.

76. Hall to Sidney and Sarah Budington, March 10, 1863, March 14, 1863, March 31, 1863, all in item 40, CFHC NMAH.

77. Hall, "Journal," April 22, 1863, item 55, CFHC NMAH. This doctor, recorded as Dr. Gay, tended to the Inuit through numerous illnesses, and Hannah wrote that he was "so kind to us always." Hannah to Sarah Budington, December 12, 1863, Ms Eb47 M1160, ICRC.

78. Hall, "Journal," April 27, 1863, May 4, 1863, item 55, CFHC NMAH.

79. Hall and Hannah to Sarah Budington, August 17, 1863, Ms Eb47 M1159, ICRC.

80. Hannah to Sarah Budington, [Summer 1863—undated letter from Nyack, NY], Ms Eb47 M1186, ICRC.

81. Hannah to Sarah Budington, August 24, 1864, Ms Eb47 M1161, ICRC.

82. Luciano, *Arranging Grief*.

83. Hall and Hannah to Sarah Budington, August 17, 1863, Ms Eb47 M1159, ICRC; Hannah to Sarah Budington, August 24, 1864, Ms Eb47 M1161, ICRC.

84. Hall, Arctic Researches, 574.

85. Mariano Aupilaarjuk and Rose Iqallijuq in Aupilaarjuk et al., *Cosmology and Shamanism*,

27–28, 128. Andrew Dialla told me about people's essences being attached to grave objects (pers. comm.).

86. Hall, "Journal," April 27, 1863, item 55, CFHC NMAH.

87. Louee Mike, in a 1989 interview with Karla Jessen Williamson, as transcribed in the Karla Jessen Williamson fonds, G91-005, Prince of Wales Heritage Centre, Yellowknife, NT. When I asked Louee for permission to use this quote, she tried to remember why she had said it decades earlier. She speculated that maybe she really wanted people to understand how much the land is a part of Inuit lives and ways of life. She added that the land and the air and the sea give Inuit a great deal of information about how things will be in the near future. For the results of Dr. Williamson's project, see Karla Jessen Williamson, "The Cultural Ecological Perspectives of Canadian Inuit: Implications for Child-Rearing and Education," (master's thesis, University of Saskatchewan, 1992).

88. Hannah to Sarah Budington, July 10 and August 24 and 27, 1864, Ms Eb47 M1164, ICRC; Hannah to Sarah Budington, August 10, 1866, item 88, CFHC NMAH.

89. J. J. Copp, "Handwritten Article about Sylvia Grinnell Little Punny," Ms Eb47 M1187, ICRC.

90. Hannah to Sarah Budington, [Summer 1863—undated letter from Nyack, NY], Ms Eb47 M1186.

91. Hannah to Sarah Budington, April 18, 1864, Ms Eb47 M1162, ICRC. Hannah's meaning is unclear to me, but she does seem to be referring to home as the Arctic.

92. Butler, "Miscellaneous typed working notes . . . ," ICRC, 34-35.

93. Michèle Therrien, pers. comm. about the concept of home and the importance of traditional dwellings for communication.

94. Matt, "You Can't Go Home Again," 479, 482–84.

95. Hall to Sidney Budington, March 10, 1863, item 40, CFHC NMAH.

96. Hall's address to the American Geographical Society, June 26, 1871, as reported in *Journal of the American Geographical Society of New York*, vol. 3 (Albany, NY: Argus Company, 1873), 406.

97. Hannah to Sarah Budington, July 10 and August 24 and 27, 1864, Ms Eb47 M1164, ICRC; Hannah to Sarah Budington, August 10, 1866, item 88, CFHC NMAH; Ipiirvik to Henry L. Brevoort, June 20, 1870, item 110, CFHC NMAH; Sarah Bolton, "Hannah and Joe: Two Faithful Little Eskimos," [1895, unknown newspaper], in MR 58, MSM.

98. Hall, "Notes for Lectures," item 46, CFHC NMAH.

99. Hannah to Sarah Budington, [Summer 1863—undated letter from Nyack, NY], Ms Eb47 M1186, ICRC.

100. Hall to Budington, June 21, 1863, item 40, CFHC NMAH.

101. Hall's note above his copy of a letter from Grinnell to Sidney Budington [June 1863], item 40, CFHC NMAH.

102. Hall, "Journal," April 15, 1863, item 55, CFHC NMAH.

103. "Lecture on 'Life among the Esquimaux,'" *Rhode Island Press*, January 10, 1863; Hall, *Arctic Researches*, 571; "The Arctic Discoveries," [unknown newspaper], November 8, 1862, item 3, CFHC NMAH.

104. Etooangat Aksayuk, interview with Margaret Nakashuk, AVC. The other man's name was Tasha. Translated from Inuktitut by Andrew Dialla.

105. Qitsualik, "Ilira," pts. 1 and 3; Kuptana, "Ilira," 7; Brody, *Other Side of Eden*, 42–44.

106. Stearns, *American Cool*, 29–33.

107. "The Polaris, The Course to Be Pursued with the Buddington Party." [Clipping, unknown newspaper and date], George E. Tyson Papers, NARA MD.

108. Loomis, *Weird and Tragic Shores*, 194–97.

109. Hall, "Death of Patrick," in item 91, CFHC NMAH.

110. John A. Macdonald to Henry Grinnell, December 13, 1869, item 110, CFHC NMAH. The first Prime Minister of Canada, Macdonald was then serving as Minister of Justice. See also Kenn Harper, "Murder at Repulse Bay: Part 2," *Nunatsiaq News*, September 14, 2007.

111. Dodge and Loomis, "Hall, Charles Francis."

112. US Navy Department, *Annual Report*, 329–30, 349–51; Loomis, *Weird and Tragic Shores*, 282, 292.

113. Hall, "Journal," September 15, 1861, item 21, CFHC NMAH (emphasis in original). See also Hall, *Arctic Researches*, 422.

114. "An Esquimau Ethnograph: Men, Women, and Manners in the Igloo Land," *St. Louis Globe Democrat*, October 23, 1880. Punctuation per original.

115. Contract naming Caladon L. Daboll as Hall's attorney, December 13, 1862, item 52, CFHC NMAH. See also Louis Agassiz (the scientist and opponent of Darwin) to Hall, [October?] 29, 1862, item 45, CFHC NMAH.

116. Hall to Grinnell, June 16, 1870, June 8, 1870, January 27, 1871; all in Coll. 8, MSM. See also Hall to Capt. and Mrs. Budington, April 15, 1863, item 40, CFHC NMAH; Hall to Secretary of the Navy George M. Roberts, October 21, 1871, reprinted in "Captain Hall and the Polaris," *Harper's Weekly*, July 5, 1873, 573. Other American explorers compared Indigenous peoples to children; see, e.g., Robinson, *Coldest Crucible*, 46.

117. Hall, *Arctic Researches*, 217, 442; Hall to Sarah Budington, December 7, 1862, item 40; Hall to Sidney Budington, January 21, 1861, item 110, both in CFHC NMAH.

118. Ancestry.com, *1860 United States Federal Census*, Cincinnati Ward 2, Hamilton, Ohio; Hall, *Arctic Researches*, xxii, xxvi (timing of trips east). His wife petitioned the government for money after his death. When she died, she had enough to put $5,000 in a trust and leave the rest to her unmarried daughter. Russell Potter, "Mercy Ann Hall," *Visions of the North*, March 13, 2012, http://visionsnorth.blogspot.ca/2012/03/mercy-ann-hall.html; "[Will of Mercy A. Hall]," Ancestry.com, *Ohio, Wills and Probate Records, 1786–1998*, vols. 80–82, 1900–1901.

119. On Hall's reliance on Inuit hunters and seamstresses see, e.g., Hall, *Arctic Researches*, 181, 217.

120. Hall, *Arctic Researches*, 482.

121. J. J. Copp, "Handwritten Article about Sylvia Grinnell Little Punny," Ms Eb47 M1187, ICRC (Grandpa Hall); J. C. Brevoort, "Joseph Ebierbing's book," 1875, MsBd Eb47, ICRC (Father Hall).

122. Hall to Grinnell, April 16, 1871, Coll. 8, MSM; Hall to Grinnell, March 4, 1863, item 40, CFHC NMAH.

123. The anthropologist Jean Briggs was not called consistently or universally by kinship terms when she was adopted by an Inuit family. Briggs, *Never in Anger*, 63–64, 302–3. See also Fienup-Riordan, "What's in a Name? Becoming a Real Person in a Yup'ik Community," in Kan, ed., *Strangers to Relatives*, 236–38; and Kan, "Editor's Introduction," in the same volume.

124. Hall to Budington, [June 2, 1861?], item 45, CFHC NMAH.

125. Hall, draft of a letter sent to R. W. Seager, "Journal," February 12, 1863, item 55, CFHC NMAH.

126. Hannah to Sarah Budington, April 1, 1864, Ms Eb47 M1161, ICRC; Hannah to Sarah Budington, August 17, 1863, Ms Eb47 M1159, ICRC.

127. Ekho and Ottokie, *Childrearing Practices*, 91.

128. Harper, *In Those Days*, 74–76.

129. Hall to Grinnell, October 13, 1870, Coll. 8, MSM. Several other letters in this collection also relate to financial transactions between Hall and Grinnell.

130. Hall to Grinnell, October 17, 1870, Coll. 8, MSM.

131. Nickerson, *Midnight to the North*, 154 (says Copp had power of attorney on their home); Kenn Harper (pers. comm. about Copp holding mortgage). Kenn notes that Hannah and Joe's house was still standing when he last visited. Surviving deeds are from the Town of Groton Land Records: Transfer from New London Savings Bank to Joseph Ebierbing, July 10, 1871, p. 775 (contains original purchase date); Transfer from Sidney O. Budington and Joseph Ebierbing, March 8, 1875, p. 167 (says land deeded to Budington in December 1870).

132. "Current Mention," *Independent Statesman* (Concord, NH), July 17, 1873, 36 (quote).

133. Butler, "Miscellaneous typed working notes . . ." ICRC, 39-40.

134. Hall to Henry Grinnell, June 5 and 8, 1870, Coll. 8, MSM. Emphasis in original.

135. Ipiirvik to Henry L. Brevoort, June 20, 1870, item 110, CFHC NMAH.

136. I have not found the contract with Barnum, but they were promised $100 per week plus expenses to work at Cutting's aquarium in Boston. Draft of a letter from Hall to William Guay, November 11, 1862, item 45, CFHC NMAH.

137. Hall and Hannah to Sarah Budington, August 17, 1863, MS Eb47 M1159, ICRC (board); Hall to Sarah Budington, October 20, 1863, item 40, CFHC NMAH ("could not spare" the money).

138. Hall to Grinnell, June 30, 1870, Coll. 8, MSM (taking money from "Joe and Hannah's fund" to pay Frank Lailer); Hall, "Expenditures of Hall," item 58, CFHC NMAH.

139. Hannah to Sarah Budington, April 18, 1864, Ms Eb47 M1162, ICRC.

140. Hall, Arctic Researches, 461.

141. Gilder, *Schwatka's Search*, 11–12.

142. On soliciting for lectures, see Wells Publishing Company to George Tyson, May 14, 1873, George E. Tyson Papers, NARA MD. On employment/work, see, e.g., Nourse, ed., *Narrative*, 444; "Esquimau Joe," *Daily Inter Ocean*, January 1, 1876; J. E. Nourse to J. J. Copp, December 14, 1876, Ms Eb47 M1183, ICRC.

143. On finances see the following in MsEb47 at ICRC: J. Morison to Hannah Ebierbing, February 13, 1875 and March 2, 1875, M1176 and M1172; J. G. Bennett to Hannah Ebierbing, September 3, 1875 and November 16, 1875, M1175 and M1173; H. Merriman to Hannah, July 19, 1876, M1180; Mrs. George W. Bailey to Hannah, September 9, 1873, M1167; Joseph Warren to Hannah, September 9, 1873, M1168; [Colonel J. Lupton] to Hannah, November 4, 1874, M1171B ($1336 and first record of $20 allowance, probably while Joe was on the *Tigress* but this is not specified). On government settlement, see "What Was Done by the Law-makers Yesterday," *Milwaukee Daily Sentinel* (Milwaukee, WI), May 23, 1874. On gifts see, e.g., J. J. Copp, "Handwritten Article about Sylvia Grinnell Little Punny," Ms Eb47 M1187, ICRC; Hall to Sidney Budington, January 18, 1863, item 40, CFHC NMAH; Colby, "Joe and Hannah," ICRC, DOC 1939-04-00A.

144. Hannah to Mr. L Brovent, [March 14, 1875], Ms Eb47 M1178, ICRC. (The letter is mistakenly identified as having a date of September 14, 1875).

145. Colby, "Tells Intimate Stories," ICRC, DOC 1939-03-00A.

146. Joamie in Ootoova et al., *Perspectives on Traditional Health*, 205, see also 1–2, same volume.

147. Hall to Grinnell, February 7, 1871, Coll. 8, MSM (whooping cough); Hannah to Sarah Budington, June 22, 1873, Ms Eb47 M1169, ICRC (unspecified illness that lasted for weeks).

148. Bradbury, "Fragmented Family," 109.

149. Hall, *Arctic Researches*, 463; Hall to Sidney Budington, January 27, 1863, item 40, CFHC NMAH (quote about "our people.") On tuberculosis and influenzal pneumonia, see Mulder and

Stuart-Harris, "Influenzal Pneumonia," 743–53; William D. Johnston, "Tuberculosis," in Kiple, ed., *Cambridge Historical Dictionary of Disease*, 336–42; Spink, *Infectious Diseases*, 219; Grygier, *Long Way from Home*, 4–6; Hays, *Epidemics and Pandemics*, 201–10.

150. On Hannah and Ipiirvik's use of *angakkuit*, see Hall, *Arctic Researches*, 345–48; Nourse, ed., *Narrative*, 248–50.

151. Ootoova et al., Perspectives on Traditional Health, 51, 148–49.

152. Hall, "Journal," February 16, 1860, item 4, CFHC NMAH.

153. Daniel, *Captain of Death*, 166.

154. Henry I. Bowditch, "Is Consumption Ever Contagious" (1864) and "Consumption in America" (1869), in *From Consumption to Tuberculosis*, 43–56 and 57–96.

155. Hannah to Sarah Budington, August 17, 1863, Ms Eb47 M1159, ICRC.

156. Aalasi Joamie in Ootoova et al., *Perspectives on Traditional Health*, 200.

157. Hall, *Arctic Researches*, 160.

158. Maddox, *Removals*, 22–23.

159. "Esquimau Joe," *Daily Inter Ocean* (Chicago, IL), January 1, 1876.

160. Hannah to Sarah, April 1, 1864, Ms Eb47 M1162, ICRC.

161. On illness and power relations, see Tester, McNicoll, and Irniq, "Writing for Our Lives," esp. 128–32.

162. On using American doctors and remedies, see, e.g., Hall, "Journal," April 15, 1863, item 55, CFHC NMAH; Hall to Grinnell, February 6, 1871, Coll. 8, MSM; Nourse, ed., *Narrative*, 254.

163. Ilisapi Ootoova and Tirisi Ijjangiaq, in Ootoova et al., *Perspectives on Traditional Health*, 20. For a nineteenth-century example of sick people requesting specific foods and the efforts to provide it, see the story recorded by James Mutch in Boas, "Eskimo of Baffin Land," 277–78.

164. Colby, "Tells Intimate Stories," DOC 1939-03-00A, ICRC.

165. Ootoova et al., *Perspectives on Traditional Health*, 51, 148–49. Tipuula Qaapik Atagutsiak notes that "as long as it was oil, it didn't matter what animal it came from" (51). On castor oil, see Nickerson, *Midnight to the North*, 160.

166. "Esquimau Joe," *Daily Inter Ocean* (Chicago, IL), January 1, 1876.

167. "Death of the Esquimaux Child," *New York Herald*, March 1, 1863.

168. Spufford, *I May Be Some Time*, 228; Valenčius, *Health of the Country*, esp. 22–37. See also Clare, "Feeling Cold," 178–80.

169. "Death of Hannah, the Wife of Esquimaux Joe," *Arizona Miner* (Prescott, AZ), February 16, 1877 (wet and changeable quote); "Esquimau Joe," *Daily Inter Ocean* (Chicago, IL), January 1, 1876; "Death of the Esquimaux Child," *New York Herald*, March 1, 1863; "The Change Too Great; Peary's Esquimaux Cannot Stand the Climate," *Portland Oregonian*, November 3, 1897; "Death of Punnie, the Esquimaux Girl," *New York Times*, March 19, 1875; "Living a Fairy Tale: Little Eskimo Boy Who Is Being Educated In New York," *Atchison Daily Globe* (Atchison, KS), January 27, 1899; Hall, "Journal," June 25, 1860, July 1, 1860 (on Kudlago and fog, shorter version in Hall, *Arctic Researches*, 40), item 12, CFHC NMAH.

170. Hall, *Arctic Researches*, 131. He cited Acts 17:26: "[God] hath made of one blood all nations of men to dwell on the whole face of the earth, and hath determined the times before appointed, and the bounds of their habitation."

171. See, e.g., David S. Jones, "Virgin Soils Revisited," in *William and Mary Quarterly* 60:4 (October 2003), 703–42; Jones, *Rationalizing Epidemics*; Daschuk, *Clearing the Plains*; Tester, McNicoll, and Tran, "Structural Violence"; Piper, "Chesterfield Inlet."

172. "Death of Hannah, the Wife of Esquimaux Joe," February 16, 1877.

173. Nourse, ed., *Narrative*, 445.

NOTES TO PAGES 75–82

174. US Navy Department, *Annual Report*, 352.

175. Italoo Enoch to "Brother and Sister," September 19, 1874, ICRC; Harper, *In Those Days*, 71–73.

176. Nourse, ed., *Narrative*, 447–48.

177. Colby, "Tells Intimate Stories," DOC 1939-03-00A, ICRC.

178. Sarah Bolton, "Hannah and Joe: Two Faithful Little Eskimos," [1895, unknown newspaper], in MR 58, MSM. This was not the first time Ipiirvik had expressed concern over traveling to the Central Arctic. In 1870, Joe dictated a letter which read, "try go King Williams Land spring . . . make me afraid . . . next time I go like a soldier." Ipiirvik to Henry L. Brevoort, June 20, 1870, item 110, CFHC NMAH.

179. Gilder, *Schwatka's Search*, 58.

180. Gilder, *Schwatka's Search*, 265–66.

Chapter Three

1. Adolphus Greely was the army's top meteorologist. He had been twice wounded in the Civil War and supervised the construction of military telegraph lines in Texas and the American West. He had been directly involved in conflicts with First Nations in 1869–70. "Record of General A. W. Greely—Personal," box 100, Greely Papers, LOC; Robinson, *Coldest Crucible*, 92.

2. Kislingbury to Greely, January 13, 1881, box 65, Greely Papers, LOC.

3. Greely, *Three Years*, 1:52–53.

4. On the IPY, see Barr, *Expeditions*. Today these IPY scientific records are being revisited to study climate change. Michael Robinson, in Public Broadcasting Corporation, "Interview with Michael Robinson." The idea of an American "colony" on Ellesmere Island had been brewing for over a decade, and in 1877 the whaling captain George Tyson brought fifteen Inuit from Cumberland Sound to Greenland as part of this venture, but he found it had been postponed and returned the Inuit to Cumberland Sound. See Dick, *Muskox Land*, 182–86; Tyson, *Cruise*. On the impact of this "colony" on Canadian sovereignty over the Arctic Archipelago, see Grant, *Polar Imperative*, 155–68.

5. Greely, *Report*, 112–14; Greely, *Three Years*, 1:85.

6. Kislingbury, "Journal," May 1, 1884, box 13, RG-27.4.5, NARA MD (this collection hereafter cited as RG-27.4.5).

7. Dick, *Muskox Land*, 26–30 (climatic shifts), 212–14 (reasons for expedition's failure); see this book for more on the human history of Ellesmere Island in the past two centuries. For Baffin trips, see, e.g., the Qillarsuaq story in the prologue of this volume, and Mary-Rousselière, *Qitdlarssuaq*, 154–55.

8. Dick, "The Fort Conger Shelters."

9. George Rice, Draft of love letter to unknown recipient [possibly Helen Bishop], box 9, Lady Franklin Bay Collection, EC.

10. Sergeant David Brainard to his mother, August 1, 1882, box 8, EC.

11. "Two Philadelphians Who Perished in the Ice of the Arctic Regions," *North American* (Philadelphia, PA), July 19, 1884. On Gardiner and his family, see also Urness, *Twenty-Five Brave Men*, 161–72; US Census of 1850, South Ward Philadelphia, County of Philadelphia, Pennsylvania, October 8, 1850, Dwelling 400, Family 422; US Census of 1880, Grant Township, Pender County, NC, June 3, 1880, Page 7, Supervisor's District 3, Enumeration District 159, Dwelling 63, Family 69.

12. Greely, *Three Years*, 1:130–32, 148; 2:65, 113, 404.

13. Gardiner, "Journal 1881–1882," March 1, 1882, box 1, EC; "Extract from Sergeant D. C.

Ralston's Diary" September 28, 1883, box 13, RG-27.4.5; Octave Pavy's copied and translated notes, February 11, 1882, box 12, RG-27.4.5; Pavy in Greely, *Report*, 326.

14. Schneider, "Journal," September 1, 5, and 11, 1881, box 69, Greely Papers, LOC; Gardiner, "Journal 1881–1882," most entries from October 9, 1881 to November 24, 1881, EC.

15. Greely, *Three Years*, 1:166; Schneider, "Account of Jens Running Away," box 3, EC.

16. Gardiner, "Journal 1881–1882," November 2, 1881, EC.

17. Markham, ed., "Arctic Expedition," 4–5.

18. Kane, U.S. Grinnell expedition, 273.

19. Hayes, *Open Polar Sea*, 201.

20. Lockwood to his father, January 9, 1882, box 4, EC.

21. Gardiner, "Journal 1881–1882," February 18, 1882, EC.

22. Brainard, *Six Came Back*, 37.

23. Greely, "The Scope and Value of Arctic Explorations," *Report of the Sixth International Geographic Congress* [n.d.], box 92, Greely Papers, LOC.

24. Brainard to his mother, August 1, 1882, EC.

25. Lockwood to his father, January 9, 1882.

26. Rice, "Journal: Sergeant Rice," June 17, 1883.

27. Gardiner, "Journal 1881–1882," November 22, 1881, EC.

28. Brainard to his mother, August 1, 1882, EC.

29. Charles Henry, "Journal Written as a Letter to the Gatter family," [entry n.d., early Dec 1881], box 3, EC.

30. Rice, "Journal: Sergeant Rice," June 7, 1882.

31. Gardiner, "Journal 1881–1882," March 10, 1882, EC. Gardiner may have read earlier exploration narratives that commented on stillness and silence. See, e.g., Loomis, "Arctic Sublime," especially the quote from William Parry on p. 102. David notes that the "sublime" was referenced in Arctic accounts into the 1880s, when it was unfashionable elsewhere, in *Arctic in the British Imagination*, 240.

32. Greely, *Three Years*, 2:208; Brainard, "Diary," October 29, 1883, in Greely, *Report*, 462.

33. Kislingbury, "Journal," August 16, 1883, RG-27.4.5.

34. "A Talk with General Greely: The Awful Silence of the Arctic Region," *Atlanta Constitution*, December 7, 1888, in box 73, Greely Papers, LOC.

35. See, e.g., Tuan, *Space and Place*, 16; Baudrillard, *America*, 6.

36. Bachelard, *La terre*, 378–85. Translated as *Earth and Reveries of Will*.

37. Baudrillard, *America*, 3, 6.

38. Baudrillard, *America*, 16. Baudrillard is referring here to the vastness of the entire American continent and its skies.

39. Brainard to his mother, August 1, 1882, EC.

40. Brainard to his mother, August 1, 1882, EC.

41. Gardiner, "Journal 1881–1882," February 28, 1882, EC.

42. Gardiner, "Journal 1881–1882," March 5, 1882, EC.

43. Brainard to his mother, August 1, 1882, EC; Greely to Henrietta Greely, November 27, 1881 and December 16, 1881, box 14, Greely Papers, LOC.

44. *Arctic Moon*, November 24, 1881, box 1, Brainard Collection 200.13-LFB, NARA MD; *Arctic Moon*, January 1, 1882, box 7, EC; Saladin d'Anglure, "Brother Moon."

45. Greely, "Our First Christmas in the Arctic," in *Book Buyer*, December 1885, 288, in box 81, Greely Papers, LOC; David Brainard to Wilkens, [n.d.], box 8, EC; Brainard to his mother, August 1, 1882, EC.

46. "Dinner Menu Christmas 1881," box 7, EC; Schneider, "Journal," December 24, 1881, box 3, EC.

47. Greely, "Journal," December 13, 1881, box 22, RG-27.4.5.

48. Greely, *Report*, 13–14; Brainard to his mother, August 1, 1882, EC; Schneider, "Journal," December 13, 1881, in Greely Papers, box 69, LOC.

49. The 1870 Danish census records 112 individuals living in sixteen households in Prøven. Only three inhabitants were Danish colonists living in European-style homes, although several of the Kalaallit (including Thorleif Frederik Christiansen) had some Danish ancestors. Dansk Demografisk Database, RA.

50. Nuttall, "The Name Never Dies," 124.

51. Vital statistics information for Jens Edvard and relatives from Upernavik Museum, "Stamtræ," http://www.upernavik.museum.gl/Files/Filer/Upernavik/Upernavimmiutoqqat/index.html.

52. Greely, "Journal" [originals], December 13, 1881, box 70, LOC.

53. Greely to Chief Signal Officer of the Army, Upernavik July 28, 1881, box 18, RG-27.4.5. The Danish colonial records record the hiring of the two men but not the reasons. "Abskrift av Assistents Journalen far Anlaget Prøven far Handelsaaret 1881/82," July 26, 1881; "Dagbøg fort ved Kolonierne Edgesminde og Upernavik samt Anlaget Godhavn i Handelsaaret 1881/82" av Volontar E. F. Myhre, July 29, 1881, both in Dagbøger af kolonibestyrere, assistenter m. fl. i Grønland, RA. See also Kopier af Nordgrønlands inspektorats korrespondance (officielt), RA, July 29, 1881.

54. Hendrik, *Memoirs*, 83. He volunteered for the first expedition after no one else did (22); he joined the second one only after they agreed he could bring his wife (36); he left on the *Polaris* when the local Danish trader was behaving angrily toward him and then asked him to go (48).

55. Bordin, "La nuit inuit," 51–52, 65–66. The dictionaries are Taamusi Qumaq's (Nunavik) and Elisapee Ootoova's (Pond Inlet/Mittimatalik).

56. Saladin d'Anglure, "Brother Moon," 189.

57. Hendrik *Memoirs*, 24. See also Anna Nungaq's recollections of being relocated as a child, as cited in Dick, *Muskox Land*, 438.

58. Census data from Dansk Demografisk Database, RA.

59. Flora, "The Lonely Un-dead," 49.

60. On *qivittut*, see Grønnow, "Blessings and Horrors" (esp. 193); Nuttall, *Arctic Homeland*, 112–14; Petersen, "Om qivittut"; Sonne and Kleivan, *Eskimos Greenland and Canada*, 21–22, Hendrik, *Memoirs*, 50–51. For older recorded *qivittoq* stories, see Sonne, *Grønlandske fortællinger*. This database contains Danish summaries and translations; stories would originally have been told (and sometimes recorded) in Greenlandic. Due to language and financial constraints I have used the Danish summaries only.

61. "Oqalugtuaq áipâ / Qivigtoq / Qivittoq," Christian Berthelsen's summary and translation from Greenlandic in Sonne, *Grønlandske fortællinger*, Document ID 173, my translation from Danish. This story is from much further south in West Greenland, from Aasiaat/Egedesminde. There is another variation from Ilulissat or Ilimanaq, closer to Kangersuatsiaq, see "Bjørnene i menneskeham," Document ID 1603. This woman went *qivittoq* in response to her father's incest.

62. Hendrik, *Memoirs*, 89–91.

63. Greely, "Our Kivigtok: An Episode of the Lady Franklin Bay Expedition," [published article in unknown source], 563, in box 73, Greely Papers, LOC.

64. Petersen, "Om qivittut."

65. Greely, *Three Years*, 1:169; Greely, "Our Kivigtok," 560.

66. Greely, "Journal 1881–1882," December 26, 1881, box 22, RG-27.4.5; "Lime Juice Club at Dutch Island Opera House, Programme, Monday Dec 26 1881," box 7, EC.

67. Greely, *Three Years*, 1:90, 164.

68. Greely, "Our Kivigtok," 561. For mentions of homesickness in letters to his wife, see Greely to Henrietta Greely, October 16, 1881, and January 3, 20, 22, and June 11, 1882, in box 14, Greely Papers, LOC.

69. Ralston, "Diary," September 30, 1883; Cross, "Journal," September 30, 1883, both in box 13, RG-27.4.3, NARA MD.

70. Henry, "Diary," October 9, 1883, box 13, RG-27.4.5.

71. Cross, "Journal," October 8, 1883, box 13, RG-27.4.5.

72. Some other expedition members appreciated the Kalaallit employees more than Greely. Gardiner said Angutisiak was "very industrious and works hard"; Kislingbury said the two men were "worth their weight in gold" (Gardiner, "Journal 1881–1882," March 15, 1882, EC; Kislingbury, "Journal," September 2, 1883, RG-27.4.5).

73. Greely, *Three Years*, 1:246–48. Sergeant Rice, who was on one of these sledge trips, said that "doubtless Jens' exhaustion was due to the greater exertions he made" (cited on 247). For an example of Jens's frequent travel, see Gardiner, "Journal 1881–1882," November 8, 1881, EC. On Greely's condemnation of Inuit endurance see also Greely, "Journal," March 18, 1884, box 12, RG-27.4.5.

74. Greely, *Three Years*, 1:245.

75. Brainard as cited in Greely, *Three Years*, 1:341.

76. Greely, "Journal," December 15, 1881, box 22, RG-27.4.5.

77. Cross, "Diary," December 13, 1881, box 3, EC.

78. Greely, *Three Years*, 1:228–29.

79. Schneider, "Account of Jens Running Away," box 3, EC.

80. Greely, "Journal," December 13, 1881, box 22, RG-27.4.5; Pavy, "Diary," December 6–8, 1881, box 12, RG-27.4.5. For more on the Kalaallit employees' reported actions before Angutisiak's flight, see Dick, *Muskox Land*, 370.

81. Biederbick, "Medical Report," in Greely, *Report*, 336; Greely, *Three Years*, 1:168.

82. Greely, "Our Kivigtok," 560, in box 73, Greely Papers, LOC (turned his back); Charles Henry, "Journal written as a letter to the Gatter family," December 13, 1881, box 3, EC (home or die).

83. For a description of the launch, see Guttridge, *Ghosts of Cape Sabine*, 47.

84. Greely, *Report*, 50.

85. Robinson, *Coldest Crucible*, 93.

86. Greely to Henrietta Greely, December 25, 1882, box 14, Greely Papers, LOC.

87. Greely, *Report*, 343.

88. Biederbick, "Medical Report," in Greely, *Report*, 336–37; Brainard to his mother, August 1, 1882, EC.

89. Greely, *Three Years*, 2:71–72.

90. Cross, "Journal," July 29, 1883, box 13, RG-27.4.5. Cross was Greely's most open antagonist, so his account may not be reliable.

91. Cross, "Journal," August 11–24, 1883; September 8, 1883.

92. Gardiner, "Diary," August 12, 1883 (cold), August 14, 1883 (bags), August 17, 1883 (rations), box 13, RG-27.4.5; Gardiner, "Journal 1881–1882," EC, February 18, 1882 (collar). On inadequate clothing and use of blankets, see Biederbick, "Medical Report," in Greely, *Report*, 335; Cross, "Diary," March 15, 1882, box 3, EC; Greely to Pavy, May 9, 1883, box 15, RG-27.4.5.

93. Ralston, "Diary," September 6, 1883 (bag), September 21, 1883 (boots), box 13, RG-27.4.5.

94. Brainard, diary entries, September 27 and 29, 1883, in Greely, *Report*, 453.

95. McClintock, *Voyage of the Fox*, 141. For another long journey from Cumberland Sound, see Hall in Davis, ed., *Narrative*, 206.

96. Jamesie Mike, interview, August 29, 2008, AVC. On trails, see Aporta, "Routes, Trails, and Tracks" and "The Trail as Home."

97. Frank B. Copley, "The Measure of Human Grit: The Struggle for the Meat," *American Magazine* 71 (1911): 254–55.

98. Greely, "Lieutenant Greely's Supplementary Report on Hygiene," in Greely, *Report*, 345; Green, "Medical report," 261.

99. Russell, *Hunger*, 124, 127.

100. Ralston, "Diary," February 4, 1884, box 13, RG-27.4.5.

101. Green, "Medical Report," 260.

102. For Elison's story, see Schneider, "This Account Is Written by the Dictation of Sergt Joseph Elison at Camp Clay, June 8, 1884," box 13, RG-27.4.5. On food for Elison, see Greely, *Report*, 81; Greely, "Supplementary Report on Hygiene," in *Report*, 346.

103. Cross, "Diary," December 21, 1883, box 13, RG-27.4.5.

104. In 1880, the medical doctor Henry Tanner disproved a common notion that humans could only survive ten days without food by completing a forty-day fast at Clarendon Hall in New York City, but many scientists believed he had cheated. Russell, *Hunger*, 53.

105. Kislingbury, "Journal," February 22, 1884, RG-27.4.5.

106. Greely, "Journal," February 17, 1884, box 12, RG-27.4.5; see also Greely, *Three Years*, 2:275.

107. Rice, "Journal," March 4, 1884, box 13, RG-27.4.5. For a physical description of Rice see "Descriptive List and Account of Pay and Clothing of George W. Rice," November 17, 1884, box 72, Greely Papers, LOC.

108. Brainard, "Diary," March 7, 1884, in Greely, *Report*, 490.

109. Israel, "Diary," March 11, 1884, box 13, RG-27.4.5.

110. Brainard, "Diary," March 16, 1884, in Greely, *Report*, 494.

111. Brainard, "Diary," March 25, 1884, in Greely, *Report*, 498.

112. Henry, "Diary," April 24 and 26, May 16, 1881, box 13, RG-27.4.5. The men connected sunlight with joy and health, but they do not seem to have advocated heliotherapy as a medical cure the way later generations would. By the 1920s heliotherapy was a popular treatment for many diseases. See Emerson, "Sunlight and Health."

113. Brainard, "Diary," April 18, 1884, in Greely, *Report*, 507.

114. Greely, *Three Years*, 2:275–76.

115. On collecting plants, a.k.a. the "flower craze," see Rice, "Journal: Sergeant Rice," July 12, 1883.

116. Taamusi Qumaq, as cited in Bordin, "La nuit inuit," 51 (my translation from French).

117. Greely to Chief Signal Officer from St John's, Newfoundland, July 17, 1884, box 14[?], Greely Papers, LOC.

118. Greely, "Journal," March 16, 1884, box 8, RG-27.4.5. Greely called the most common birds dovekies but the described size (over a pound of meat per bird) suggests they must have been black guillemots. See Weslawski and Legezynska, "Chances for Arctic Survival," 374.

119. Ralston, "Diary," April 14, 1884, box 13, RG-27.4.5.

120. Henry, "Diary," May 18, 1884, box 13, RG-27.4.5.

121. Greely, *Three Years*, 2:290. Long, "An Arctic Bear Hunt," in Kersting, ed., *White World*, 103–8. Here Long portrays himself as the hero and leader, with Angutisiak as a hindrance.

122. Henry, "Diary," April 11, 1884, box 8, RG-27.4.5.
123. Brainard, "Diary," April 12, 1884, in Greely, *Report*, 505.
124. Israel, "Diary," April 13, 1884, box 13, RG-27.4.5.
125. Greely, *Three Years*, 1:355.
126. Kislingbury, "Journal," August 14, 1883, box 13, RG-27.4.5 (relish); Brainard, "Diary," September 1, 1883, in Greely, *Report*, 443 (blood); Ralston, "Diary," September 4, 1883, box 13, RG-27.4.5.
127. Greely, *Three Years*, 2:269, 270, 275, 278–79.
128. On extra rations for other party members, see Brainard, "Diary," April 14, 1884, in Greely, *Report*, 506. On extra rations for Greely, see Biederbick, "Diary," April 15, 1884, box 13, RG-27.5.4; Greely, *Three Years*, 2:291. Greely also reported that he did virtually no physical work that last year; see Greely, "Supplementary Report on Hygiene," in Greely, *Report*, 346. Dick noted Greely's extra allotment and drew a similar conclusion in *Muskox Land*, 205–6.
129. Greely, *Three Years*, 2:286. Love letters in box 7, Lady Franklin Bay Collection, EC.
130. Brainard, "Diary," April 27, 1884, in Greely, *Report*, 510.
131. Schneider, "Journal [typescript]," April 27, 1884, box 72, LOC; Greely, "Journal," April 28, 1884, box 12, RG-27.4.5; Octave Pavy, "Notes," April 24, 1884, box 12, RG-27.4.5.
132. Henry, "Diary," April 29, 1884, box 13, RG-27.4.5.
133. Brainard, "Diary," April 28, 1884, in Greely, *Report*, 511; Octave Pavy, "Notes [translated typescript]," March 23, 1884, box 19, RG-27.4.5.
134. Brainard, "Diary," April 29, 1884, in Greely, *Report*, 511–12; Greely, *Three Years*, 2:299–300.
135. Henry, "Diary," April 29, 1884, box 13, RG-27.4.5.
136. Schneider, "Journal [typescript]," April 29, 1884, box 72, LOC.
137. Brainard, "Diary," April 29 1884, in Greely, *Report*, 511–12; Gardiner, "Diary," April 29, 1884, box 13, RG-27.4.5.
138. Henry, "Diary," May 4, 1884, box 13, RG-27.4.5; Gardiner, "Diary," May 3, 1884, box 13, RG-27.4.5.
139. Greely to Chief Signal Officer (draft letter), December 11, 1884, box 66, Greely Papers, LOC. Another piece of correspondence reveals that a draft of £3 sterling (approximately $14.60 in 1885 USD, see "Measuring Worth," http://www.measuringworth.com/exchange/) was sent to Copenhagen in December 1885 for Jens Edvard's family. Whether this was an installment or the only payment remains unclear. (Hugo Hörring, Direktoratet for Den Kongelige grønlandske Handel, Copenhagen, to Greely, April 18, 1885, box 14[?], Greely Papers, LOC).
140. The Danish record notes only his Christian names. Danish genealogical data for Jens Edvard Frederik Simon Thorleifsen (son of Jens's brother Jakob Severin Thorleif Jeremias) on Upernavik Museum, "Stamtræ," http://www.upernavik.museum.gl/Files/Filer/Upernavik/Upernavimmiutoqqat/index.html.
141. Brainard, "Diary," June 21, 1884, in Greely, *Report*, 529.
142. Biederbick, "Diary," June 12, 1884, box 13, RG-27.4.5; Greely, *Three Years*, 2:292, 324–25; Brainard, "Diary," June 12, 1884, in Greely, *Report*, 526.
143. Green, "Medical Report," 261; Schneider, "Diary," June 16, 1884, box 13, RG-27.4.5.
144. Brainard, "Diary," May 21, 1884, in Greely, *Report*, 518; Greely, *Report*, 88. They pitched the wall tent on May 22.
145. "Rescued," *New York Herald*, July 18, 1884.
146. "Schley's Story: Some New and Interesting Facts Relating to the Finding of the Greely Expedition," in scrapbook in folder labeled "Personal Items 1881–1905," box 1, Greely Papers, LOC; Green, "Medical Report," 262.

147. Riffenburg, *Myth of the Explorer*, 105–6.

148. "The Extreme North: Sergeant Brainerd's [sic] Interesting Story of His Trip There," in scrapbook in box 73, Greely Papers, LOC (food and "sea flies"); Kislingbury, "Journal," April 30, 1884, RG-27.4.5 (corn quote). They still had a small amount of tea remaining, but given the negligible calories in tea, I consider them to have been living—or dying—off the land.

149. Henry, "Diary," April 25, 1884, box 13, RG-27.4.5.

150. Brainard, "Diary," March 16, 1884, in Greely, *Report*, 494.

151. Ralston, "Diary," April 26, 1883, box 13, RG-27.4.5.

152. Charles H. Harlow, "Greely at Cape Sabine," 87, in box 80, Greely Papers, LOC (hereafter cited as Harlow).

153. Brainard, "Diary," May 20, 1884, in Greely, *Report*, 518; Sergeant Francis Long, "The Sufferings of the Greely Polar Expedition, Now First Told for the Sunday World," in box 8, EC.

154. Weslawski and Legezynska, "Chances for Arctic Survival," 375.

155. Greely, *Report*, 91.

156. Schneider, "Journal [typescript]," May 27, 1884, box 72, Greely Papers, LOC.

157. Brainard, diary entries in Greely, *Report*, 517, 527–29; Greely, "Lieutenant Greely's Supplementary Report on Hygiene," in Greely, *Report*, 348.

158. Rice, "Diary," March 2, 1884, box 13, RG-27.4.5.

159. "Sergeant Connell, A Talk with the Survivor of the Greely Expedition," May 31, 1885, in box 73, Greely Papers, LOC.

160. Brainard, "Diary," April 23, 1884, in Greely, *Report*, 509.

161. Schneider, "Journal [typescript]," May 15, 1884, box 72, Greely Papers, LOC.

162. Henry, "Diary," May 15, 1884, box 13, RG-27.4.5.

163. Greely, "Supplementary Report on Hygiene," in Greely, *Report*, 348.

164. Brainard, "Diary," April 6, 1884, in Greely, *Report*, 502.

165. Brainard, as reported in Harlow, 89.

166. Brainard, "Diary," April 6, 1884, in Greely, *Report*, 502.

167. Brainard, as reported in Harlow, 90.

168. Weslawski and Legezynska, "Chances for Arctic Survival."

169. "Horrors of Cape Sabine," *New York Times*, August 12, 1884.

170. For a survey of headlines see Riffenburgh, *Myth of the Explorer*, 125–28.

171. "Pulpit Gleams," *Christian Union*, August 28, 1884; "Cannibalism in the Arctic," *Washington Post*, August 13, 1884; "Gen. Hazen's Admissions," *New York Times*, August 13, 1884; "Cannibalism," *Medical and Surgical Reporter*, September 20, 1884; "New Tale of Arctic Horror," *Indianapolis Journal*, August 16, 1884[?], in box 95, Greely Papers, LOC.

172. Andrew Dialla, pers. comm. See also Qitsualik, "Cannibal."

173. Trott, "Cannibalism Theme," 27.

174. Saullu Nakashuk, interview, September 13, 2008, trans. Andrew Dialla (in author's possession).

175. For more analysis of the press coverage and of how the survivors retold their story as one of masculinity, self-sacrifice, and will, see Robinson, *Coldest Crucible*, 96–106.

176. Guttridge, *Ghosts of Cape Sabine*, 307–8.

177. U.S. Army Center of Military History, "Medal of Honor Recipients."

178. US Census of 1900, New York State, Kings County, City of New York, Borough of Brooklyn, Ward 28, Supervisor's district 2, Enumeration district 503, Sheet 26 (daughter Sabine); Francis Long, naturalization certificate, County Court, Kings County, NY, 10 Oct 1888, L-520. Long was on the Baldwin-Ziegler expedition of 1901–2. See also "Greely Arctic Explorers: Eight Years

Since They Were Brought Back Dying," [clipping without provenance], box 73, Greely Papers, LOC.

179. Potter, *Arctic Spectacles*, 136, 201, plate 10; Riffenburgh, *Myth of the Explorer*, 103–9, 125–28, 198–99; Robinson, *Coldest Crucible*, 138. For a photo of the cyclorama, see Library of Congress, "[Diorama of the Greely Expedition . . .]," https://www.loc.gov/item/2006675167/.

180. Shook Collier [?] of Union Square Theater to Greely, August 12, 1884, box 66, Greely Papers, LOC.

181. "Sergeant Connell, A Talk with the Survivor of the Greely Expedition," May 31, 1885, in box 73, Greely Papers, LOC.

182. For recent works, see, e.g., Guttridge, *Ghosts of Cape Sabine*; Todd, *Abandoned*; and the PBS television series *American Experience*, "The Greely Expedition," season 23, episode 5.

Chapter Four

1. All quotes and information from Etooangat in this chapter are, unless otherwise noted, from Andrew Dialla's translations of Etooangat's 1994 interview with Margaret Nakashuk (née Karpik). The original recordings are on file at AVC.

2. For more on Indigenous encounters with newcomers in North America, see, e.g., Kupperman, *Indians and English*; Clendinnen, *Ambivalent Conquests*; Richter, *Facing East*; Apphia Agalakti Awa, in Wachowich et al., *Saqiyuq*, 119. For a parallel story of an Inuk receiving a watch as a gift and repurposing it, see Faulkner, *Eighteen Months*, 241.

3. For more on my use of oral history, please see the methodological essay in this volume.

4. See, e.g., Nuttall, "Anticipation"; Cameron, Mearns, and McGrath, "Translating Climate Change," 278.

5. There are several records of Qatsu's story, in which details sometimes differ. I have seen them in: Qatsu in Arnaktauyok, ed., *Stories from Pangnirtung*, 81–82; Qatsu in Bennett, "Whalers, Missionaries, and Inuit," 61; Qatsu Evic, interview with Jaypeetee Akpalialuk, 1984, G-1985-007, NTA; and a short unidentified typescript entitled "Relations between White and Inuit/Alcohol and Death" that Andrew Dialla took a picture of (provenance unknown). Missionary account in Fleming, *Perils of the Polar Pack*, 96–98. The nearby ship was the Norwegian vessel *Heimdal*.

6. Andrew Dialla told it to me. I have also seen it told on Facebook.

7. On Inuit children's fear of Qallunaat, see, e.g., Elisapee Ishulutaq on William Sivutiksaq Duval, interview, September 13, 2008, AVC.

8. Scottish captain William Penny first erected houses on Qikiqtat in 1857; the American crews of the *Daniel Webster* and *Black Eagle* built stations on Qikiqtat and Uummannarjuaq respectively in 1860. Goldring, "Southeast Baffin Island," 4.12–4.13, 4.17–4.19, 4.29–4.30. See also Goldring, "Inuit Economic Responses," 163–65; Stevenson, *Cultural Persistence*, 90–93. On trading nonwhale products see also the Accounts of the *Helen F.* and *Nile* (IMA 795), NBWM.

9. For a particularly dangerous journey, see Hantzsch, *My Life*. On disease at Qikiqtat, see Keenleyside, "Euro-American Whaling."

10. Markosie Pitseolak, 1973 interview in possession of Daisy Dialla, trans. Andrew Dialla (quotes); Pitseolak, *Markosie Pitseolak's Real Life Stories*; Pitseolak in Arnaktauyok, ed., *Stories from Pangnirtung*, 24–25. I am not sure when the Saturday tradition began; in the 1870s George Tyson issued rations every Monday. Tyson, *Cruise*, 70.

11. Nowyook, interview with Marc Stevenson, 1984 [?], G-1985-007-0012, NTA. Some Inuit far from the stations reportedly began using bows and arrows again. "The Esquimaux and Harry Lauder," *Dundee Advertiser*, October 13, 1921.

12. Malaya Akulukjuk in Arnaktauyok, ed., *Stories from Pangnirtung*, 75.

13. Cardno, *Whaler's Tale*, 55–67, quote on 60. In a more severe case of homesickness, an inquest in 1907 concluded that the Qallunaaq agent at Qikiqtat that year, a Mr. W. F. Milne, "shot himself with a gun in a moment of despair, caused by nostalgia, as shown by some letters written by him ... [and] left on his desk." Bernier, *Report*, 73.

14. Pauloosie Angmarlik, interview with Margaret Nakashuk, AVC. For a similar discussion of learning to hunt, see Pitseolak, *Markosie Pitseolak's Real Life Stories*, and Pitseolak in Arnaktauyok, ed., *Stories from Pangnirtung*, 18.

15. Etooangat Aksayuk, interview with Jaypeetee Akpalialuk, 1984, G-1985-007, NTA. These are two of the many games he describes here.

16. Etooangat Aksayuk, interview with Marc Stevenson, 1984 [?], G-1985-007-0007, NTA; Evie Anilniliak, interview, September 3, 2008, trans. Andrew Dialla, in author's possession (windows).

17. Boas, *Arctic Expedition*, 27; Etooangat Aksayuk, interview with Jaypeetee Akpalialuk, 1984, G-1985-007, NTA; Qatsu Evic, interview with Jaypeetee Akpalialuk, 1984, G-1985-007, NTA.

18. The terror and surprise Qatsu felt at being told to marry were not uncommon for young Inuit women in her era. See Apphia Agalakti Awa in Wachowich et al., *Saqiyuq*, 36–42.

19. Qatsu Evic, interview with Marc Stevenson, 1984 [?], G-1985-007-0013, NTA.

20. Ottokie, as cited in Ekho and Ottokie, *Childrearing Practices*, 26.

21. Rosee Veevee, interview, September 11, 2008, AVC.

22. Eve Nooeyout, as cited in Laugrand et al., eds., *Apostle to the Inuit*, 213. See also Edmund Peck's report of the event in the same volume, 224–28. Nuijaut is her original Inuktitut name, not a surname. Eve, or Evie, is her baptismal name.

23. Laugrand et al., eds., *Apostle to the Inuit*, 6 (first church), 143–45 (first baptisms), 6–27 (history of the missions). One of the first converts was Immukke/Mary, the grandmother of recently deceased Elder Evie Anilniliak; Evie slept with Immukke as a child. Evie Anilniliak, interview, September 3, 2008, trans. Andrew Dialla (in author's possession)

24. Peck as cited in Laugrand et al., eds., *Apostle to the Inuit*, 213.

25. Trott, "'Reading' Conversion Narratives." These types of prayers remain common today. For another early Christian Inuit example, see Hendrik, *Memoirs*, 68. On Nuijaut's knowledge of pre-Christian beliefs, see Laugrand et al., eds., *Apostle to the Inuit*, 294–96 and chap. 14.

26. Qatsu Evic, as cited in Bennett, "Whalers, Missionaries, and Inuit," 64 (turn to God quote); Katsoo Eevic in Arnaktauyok, ed., *Stories from Pangnirtung*, 78 (greater person quote).

27. Daisy Dialla, interview, September 15, 2008, AVC.

28. Katsoo Eevic in Arnaktauyok, ed., *Stories from Pangnirtung*, 79. See Laugrand and Oosten, *Inuit Shamanism and Christianity*, 134–35, on not using the term "deity" or "goddess" for Sedna.

29. Daisy Dialla, interview, September 15, 2008, AVC.

30. There is no record of what year the Sedna clothing was made. I am guessing no later than 1902, because Aasivak's conversion reportedly took place before anyone else had "turned to God," and the Sedna clothing was made later that same year. But it could be ca. 1900–1903, as Qatsu was born in 1899 and was on her mother's back during her conversion experience.

31. Daisy Dialla, interview, September 15, 2008, AVC; Trott, "Grave Disappointments," 7–11; Laugrand et al., eds., *Apostle to the Inuit*, 25–27. Trott writes that in some versions of this story, Angmarlik refused to sleep with his wife. Reverend Bilby witnessed a division of the larger community into separate camps in spring 1903, perhaps the one alluded to by Daisy Dialla above.

32. Trott, "Grave Disappointments," 6–9.

33. Etooangat was a child "during the age when my mother was still holding my hand to take

me places" when the ship *CGS Arctic* visited for the first time (1907), but he could remember quite a bit about it. He explained to Margaret that this was how the Canadian government assigned him a birthdate of 1901 for the purposes of collecting a pension. Etooangat Aksayuk, interview with Margaret Nakashuk, 1994, AVC. He died in 1996; see Müller-Wille, ed., *Franz Boas*, xii. Qatsu Evic was born in December 1899 and remembered Etooangat being born in the spring, "the time a small ship was stuck in the ice down at Kikitaet [Qikiqtat]," which may refer to the *Heimdal* in 1905. Katsoo Evic in Arnaktauyok, ed., *Stories from Pangnirtung*, 81.

34. Etooangat Aksayuk, interview with Jaypeetee Akpalialuk, 1984, G-1985-007, NTA.

35. Laugrand et al., eds., *Apostle to the Inuit*, 111, 133.

36. Etooangat, as cited in Laugrand and Oosten, eds., *Inuit Shamanism and Christianity*, 47–48.

37. Andrew Dialla, pers. comm.

38. Bordin, "La nuit inuit," 58–59.

39. On shamanism and Christianity, see Oosten and Laugrand, "*Qaujimajatuqangit*," esp. 37; Laugrand and Oosten, *Inuit Shamanism and Christianity*.

40. Description of season is from Hantzsch, *My Life*, 218. The start of floe edge whaling depended on ice and weather conditions. The timing in the text is from Etooangat, and Andrew Dialla estimated the calendar months. Pauloosie Veevee said March, and so did A. P. Low in 1903. Markosie Pitseolak said May. Franz Boas said that the main time for floe edge hunting was historically May and June, but that the hunt started as soon as the young seal hunt was over. See Etooangat Aksayuk, interview with Margaret Nakashuk, AVC; Pauloosie Veevee, interview, August 28, 2008, AVC; Low, *Report*, 9; Pitseolak, *Markosie Pitseolak's Real Life Stories*; Boas, *Arctic Expedition*, 22. New clothing from Qatsu Evic, interview with Jaypeetee Akpalialuk, 1984, G-1985-007, NTA.

41. Unless otherwise noted, information about this hunt is from Pauloosie Veevee, interview, August 28, 2008, AVC.

42. Inuusiq Nashalik, interview, August 27, 2008, AVC. His father, Attagoyuk, was a rope handler.

43. Inuusiq told a story of a whale pulling a boat onto the floe edge; this may have been the same whale or a different one. Inuusiq Nashalik, interview, August 27, 2008, AVC.

44. Pauloosie Veevee, interview, August 28, 2008, AVC.

45. Elisapee Ishulutaq, interview, September 13, 2008, AVC.

46. Pauloosie Veevee, interview, August 28, 2008, AVC; see also Cole and Müller-Wille, "Franz Boas' Expedition," 47.

47. Qatsu Evic, interview with Jaypeetee Akpalialuk, 1984, G-1985-007, NTA. Unirsagaaq is written as "Uniukshagak" in this interview but I assume it was the same person. I can't be sure it was the same whale.

48. Markosie Pitseolak, 1973 interview in possession of Daisy Dialla, trans. Andrew Dialla. See also Rosee Veevee, interview, September 11, 2008, AVC.

49. Pitseolak, *Markosie Pitseolak's Real Life Stories*.

50. Nowyook, interview with Jaypeetee Akpalialuk, 1984, G-1985-007, NTA.

51. Paulosie Angmarlik went to the floe with his adoptive father Angmarlik. Paulosie Angmarlik, interview with Margaret Nakashuk, AVC.

52. Stevenson, *Cultural Persistence*, 91–93.

53. Etooangat Aksayuk, interview with Margaret Nakashuk, AVC; Qatsu Evic, interview with Marc Stevenson, August 24–25, 1984 [?], G-1985-007-0013, NTA. Etooangat said one or more old men who were no longer whaling would often go as well. Koodloo Pitseolak said that on

Uummannarjuaq it was mostly the young boys there who hunted seals; she had never been a hunter although some women were. Koodloo Pitseolak, interview with Marc Stevenson, 1984 [?], G-1985-007-0010, NTA. On gender flexibility and ambivalence, see Trott, "The Gender of the Bear," 94–96.

54. Etooangat Aksayuk, interview with Margaret Nakashuk, AVC (returning when ice broke up); Daisy Dialla, interview, September 15, 2008, AVC; Jamesie Mike, interview, August 29, 2008, AVC (returning if they caught a whale); Inuusiq Nashalik, interview, August 27, 2008, AVC (quote from Andrew).

55. T. H. Tredgold to Mrs. Stewart, August 5, 1924, in T. H. Tredgold fonds, LAC.

56. Inuusiq Nashalik, interview, August 27, 2008, AVC.

57. See table in Stevenson, *Cultural Persistence*, 97. Pangnirtung Post records provide limited descriptions; see Pangnirtung Post, 1921–39, B.455/a/1–8 and B.455/a/8–15, HBCA. On earlier hunts, see Outerbridge, "My Voyage to Cumberland Inlet"; Wakeham, "Report of the Expedition," 73.

58. Stevenson, "Kekerten—Preliminary Archaeology," 132.

59. On motorized boats, see Peteroosie Karpik, interview, September 8, 2008, AVC. Many whaleboats continued to be in use well into the mid-twentieth century. Andrew Dialla, pers. comm. (remembered seeing Kanajuq's boat); George Anderson, "A Whale Is Killed," 21.

60. Elisapee Ishulutaq, interview, September 13, 2008, AVC.

61. Evie Anilniliak, interview, September 3, 2008, trans. Andrew Dialla (in author's possession). Inuusiq Nashalik also spoke about the *majja*, in his interview, August 27, 2008, AVC.

62. Evie Anilniliak, interview, September 3, 2008, trans. Andrew Dialla (in author's possession).

63. Inuusiq Nashalik, interview, August 27, 2008, AVC.

64. Jamesie Mike, interview, August 29, 2008, AVC. See also the Inuit Heritage Trust draft map "26J—Isuittuq," available online at http://ihti.ca/eng/place-names/pn-seri.html.

65. Qatsu Evic, interview with Marc Stevenson, 1984 [?], G-1985-007-0014, NTA; Qatsu Evic, interview with Jaypeetee Akpalialuk, 1984, G-1985-007, NTA. Qatsu refers to the whaling stations as "home" elsewhere as well. This trip was not the standard *aujaq* hunt, as the group had sleds and traveled over the snow and ice.

66. Qatsu Evic, interview with Jaypeetee Akpalialuk, 1984, G-1985-007, NTA.

67. Andrew Dialla, pers. comm. Joanasie Dialla was his father.

68. Charles Sampson, Sampson circular, September 30, 1899, Church Missionary Society fonds, C.1/0 item 1899-104, LAC. I assume this refers to Qatsu's grandparents but no names are given, so I am not sure. Thanks to Phil Goldring for drawing my attention to this source in *Southeast Baffin*, 7.22–7.23.

69. Elisapee Ishulutaq, interview, September 13, 2008, AVC; Stenton, "Caribou Population Dynamics," 21–25.

70. On Nettilling, see Hantzsch, *My Life*, 197, 214–18 (quote from 214); Nowyook, interview with Marc Stevenson, 1984[?], G-1985-007-0010, NTA; James Mutch's collections of stories in Boas, "Eskimo of Baffin Land," 135–37, 173, 236–37, 278; Boas, *Arctic Expedition*, 26. On Inuit trails, see, e.g., Aporta, "Routes, Trails, and Tracks."

71. Low, *Report*, 155; Etooangat Aksayuk, interview with Margaret Nakashuk, AVC. Qatsu said the skin boats were before her time; see Qatsu Evic, interview with Jaypeetee Akpalialuk, 1984, G-1985-007, NTA.

72. Andrew Dialla said his father had wood (pers. comm.); Hantzsch noted whale bones in *My Life*, 197.

73. Inuusiq Nashalik, interview, September 15, 2008, AVC.

74. Inuusiq Nashalik, interview, September 15, 2008, AVC; Andrew Dialla, pers. comm.

75. Boas, *Arctic Expedition*, 26. There is a very similar description in Kumlien, *Contributions*, 18.

76. Andrew Dialla, pers. comm. On rules around caribou skins, see also Laugrand and Oosten, *Hunters, Predators and Prey*, 247–50.

77. Stenton, "Caribou Population Dynamics," 19–20. See also Stenton, "Adaptive Significance."

78. Etooangat Aksayuk, interview with Margaret Nakashuk, AVC (winter hunt); Qatsu Evic, interview with Jaypeetee Akpalialuk, 1984, G-1985-007, NTA (told to leave). On caribou policy, see Sandlos, *Hunters at the Margin*, pt. 3, esp. 180, 189–91; Stenton, "Caribou Population Dynamics," 25; Kulchyski and Tester, *Kiumajut*; Loo, "Political Animals"; Natasha Thorpe, Kitikmeok Elders, et al., *Thunder on the Tundra*; Canada, *Q Book*, 274–76. For a humorous critique of *The Q Book*, see Mark Sandiford's film *Qallunaat!*

79. On family allowance, see the following thematic studies by the Qikiqtani Inuit Association: *Illinniarniq*, 38–39; *Nuutauniq*, 34–35; *Paliisikkut*, 32 (also threats of being banned from hospital). On sled dogs, see Qikiqtani Inuit Association, *Qimmiliriniq*, and pp. 11 and 54 specifically in relation to caribou hunting. On regulations, see Sandlos, *Hunters at the Margins*, esp. 208. The continuation of the hunt after resettlement is from Andrew Dialla, pers. comm.

80. Andrew Dialla, pers. comm.; Inuusiq Nashalik, interview, September 15, 2008, AVC.

81. Pauloosie Angmarlik, interview with Margaret Nakashuk, 1994, AVC.

82. Inuusiq Nashalik, interviews, August 27 and September 15, 2008 (quote), AVC.

83. For some recent newspaper coverage, see "Latest Nunavut Survey Shows Baffin Caribou Decimated," *Nunatsiaq News*, May 30, 2014; "Happy New Year: Nunavut Bans Caribou Hunting on Baffin," *Nunatsiaq News*, December 20, 2014; "Nunavut Groups Suggest Modest Quotas Replace Baffin Caribou Ban," *Nunatsiaq News*, March 9, 2015; "Baffin Caribou Moratorium Replaced by Modest Harvest," *Nunatsiaq News*, August 26, 2015; "Nunavut Hunters Have Used Fewer Than Half of Baffin Caribou Tags," *Nunatsiaq News*, February 25, 2016.

84. Thanks to Andrew Dialla for sharing his own extensive research, family photographs, and newspaper clippings about John Taylor and the *Easonian*.

85. These details are from Etooangat Aksayuk, as cited in Eber, *When the Whalers*, 162; Jamesie Mike, interview, August 29, 2008, AVC; Clark, *Last of the Whaling Captains*, 156–57; Harper, "The Burning of the Easonian," 31. The accounts of the fire complement each other: Etooangat said some kind of spark ignited the grease in the engine room; Jamesie had heard from his mother that it was the generator; Clark said it was spark plugs. On engine troubles the previous year, see "Arctic Ice and Blizzard, Dundee Vessel's Perilous Journey," *Dundee Advertiser*, October 12, 1921.

86. Hantzsch, *My Life*, 218.

87. He reportedly worked on the ships *Active*, *Morning*, and *Scotia*. Clark, *Last of the Whaling Captains*, 156.

88. "Ship Fire That Ended an Era," *Dundee Telegraph*, June 25, 1966; "The Esquimaux and Harry Lauder," *Dundee Advertiser*, October 13, 1921; "From Dundee to Trade with the Eskimos," *Dundee Evening Telegraph*, October 2, 1954.

89. "Ship Fire That Ended an Era," *Dundee Telegraph*, June 25, 1966; Clark, *Last of the Whaling Captains*, 158.

90. Etooangat Akshayuk, interview with Marc Stevenson, 1984 [?], G-1985-007-0007, NTA.

91. The late Elder Saullu Nakashuk was born at the same time as a whale was brought in,

NOTES TO PAGES 136–144

and she received a blanket from a Royal Canadian Mounted Police (RCMP) officer at her birth, which dates this hunt to after the police post was established in 1923. Saullu Nakashuk, interview, September 14, 2008, trans. Andrew Dialla (in author's possession).

92. The late interpreter Eric Joamie suggested the term *nunaliit* and the translation "communities" to me in Pangnirtung; the Qikiqtani Truth Commission recommended the term *ilagiit nunagivaktangat*, with the translation "a place used regularly or seasonally by Inuit for hunting, harvesting, or gathering." Qikiqtani Inuit Association, *Achieving Saimaqatigiingniq*, 14. On grass, see Stevenson, "Kekerten—Preliminary Archaeology," 105.

93. Pauloosie Veevee, interview, August 28, 2008, AVC.

94. Stevenson, *Cultural Persistence*, 93–100; Goldring, "Inuit Economic Responses," 166–72 (Kinnes selling up on 169; trapping when needed on 171). For Inuit complaints about the early HBC, see, e.g., Peter Tulugajuak, "A Letter of Complaint, 1922," as cited in Petrone, *Northern Voices*, 125–27; Elaiyah Keenainak as cited in von Finckelstein, ed., *Nuvisavik*, 28–29; Peteroosie Karpik, interview, September 8, 2008, AVC. On the link between the HBC and the need for relief, see Tester and Kulchyski, *Tammarniit*, 20.

95. Pauloosie Veevee, interview, August 28, 2008, AVC.

96. Daisy Dialla, interview, September 15, 2008, AVC. See also Bennett, "Whalers, Missionaries, and Inuit," 70.

97. Hantzsch, *My Life*, 39.

98. Jamesie Mike, interview, August 29, 2008, AVC.

99. Daisy Dialla, interview, September 15, 2008, AVC.

100. Daisy Dialla, interview, September 15, 2008, AVC; Jamesie Mike, interview, August 29, 2008 (wood for harpoons and copper spikes for harpoons), AVC; Pauloosie Veevee, interview, August 28, 2008 (nails into bullets), AVC. On the *Ernest William*'s wreck in 1913, see Eber, *When the Whalers*, 161; "Stirring Story of the Sea," *Northern Advocate* (New Zealand), January 5, 1914.

101. Daisy Dialla, interview, September 15, 2008, AVC.

102. Daisy Dialla, interview, September 15, 2008, AVC.

103. *Ancestors in the Attic* (season 2, episode 2027); Andrew Dialla, pers. comm.

104. Unless otherwise noted, all parts of Etooangat's famine story are from Etooangat Aksayuk to Margaret Nakashuk (née Karpik), trans. Andrew Dialla, 1994, AVC.

105. Etooangat Aksayuk, interview with Jaypeetee Akpalialuk, 1984, G-1985-007, NTA.

106. George Tyson records a wooden whaleboat being badly damaged in such conditions in Tyson, *Cruise*, 41, 46.

107. Goldring, *Inuit Economic Responses*, 159; Ross, *Distant and Unsurveyed*, 111, 120–23 (connects this to freeze-up and island location); Hantzsch, *My Life*, 43; Milwood 1867–68, January 1868, NBWM.

108. Julian Bilby to Rev. Baring-Gould, December 1905. Letter transcription from Chris Trott.

109. Goldring, "Inuit Economic Responses," 167; Etooangat Aksayuk in Arnaktauyok, ed., *Stories from Pangnirtung*, 42–43.

110. Etooangat Aksayuk, interview with Jaypeetee Akpalialuk, 1984, G-1985-007, NTA. For other stories of starvation written down by Cumberland Sound whaling captain James Mutch in the early twentieth century, see Boas, "Eskimo of Baffin Land," 275–79.

111. Daisy Dialla, interview, September 15, 2008, AVC.

112. Stevenson, *Cultural Persistence*, 98.

113. Saullu Nakashuk, interview, September 14, 2008, trans. Andrew Dialla (in author's possession). On hunger in Inuit societies, see Laugrand, Oosten, and Serkoak, "The Saddest Time of

My Life"; Loo, "Hope in the Barrenlands," in Bocking and Martin, eds., *Ice Blink*. For Qallunaat records of hunger in this region after whaling, see, e.g., Florence Hirst journal, July 17, 1938, ACC; Starvation—Baffin Land—Easter Bay and Exeter Sound, RG-85-C-1-a 9613 896, LAC; see also Hastrup, "Hunger"; Black-Rogers, "Varieties of 'Starving.'"

114. Pauloosie Veevee, interview, August 28, 2008, AVC.

115. Elisapee Ishulutaq, interview, September 13, 2008, AVC; Etooangat Akshayuk, interview with Marc Stevenson, 1984[?], G-1985-007-0004, NTA.

116. Saullu Nakashuk, in Partridge, ed., *Niurrutiqarniq*, 103. On craving bowhead, see also Nunavut Wildlife Management Board, *Final Report*, 56–57.

117. Inuusiq Nashalik, interview, September 15, 2008, AVC. To prepare the sleds for travel, hunters smeared blood on the jawbone runners as an adhesive, then used a special caribou-skin tool to sponge on a thin coating of water, which froze to ice.

118. John Macdonald, "Piugaattuk, Noah," in *Encyclopedia of the Arctic*, 1646. On regulations, see Finley, "Natural History and Conservation," 66; Mitchell and Reeves, "Factors Affecting Abundance," 67–70. Nunavut Wildlife Management Board, *Final Report*, 10, concluded that many Inuit in the Canadian Eastern Arctic believed they were banned from harvesting bowhead whales, even during periods when they were not.

119. For an environmentalist viewpoint, see Reilly et al., *Balaena mysticetus*.

120. Pauloosie Veevee, interview, August 28, 2008, AVC; Jamesie Mike, interview, August 29, 2008, AVC.

121. Billy Arnaquq, as cited in Stuckenberger, *Thin Ice*, 38. Used with permission from Billy.

122. Etooangat's story is from his interview with Margaret Nakashuk, 1994, trans. Andrew Dialla, AVC. On stories about encounters with nonhuman beings, see also Laugrand and Oosten, *Inuit Shamanism and Christianity*, 168–69.

123. Some information on *qalupaliit* is from Andrew Dialla, pers. comm. For more on *qalupaliit* (known as *qallupilluit* in North Baffin and Kivalliq regions), see Laugrand and Oosten, *Sea Woman*, 113–15; Kappianaq in *Travelling and Surviving*, 75–76; Lucassie Nutaraaluk in Aupilaarjuk et al., *Perspectives on Traditonal Law*, 183–84.

124. Andrew Dialla, pers. comm. Inuusiq Nashalik had heard them, interview, September 15, 2008, AVC.

125. Macfarlane, *Landmarks*, 71. He says he learned this from Nan Shepherd's book *The Living Mountain*.

Epilogue

1. Qikiqtani Inuit Association, *Achieving Saimaqatigiingniq*, 17.

2. The information on families returning to their land after resettlement is from Andrew Dialla, pers. comm.

3. Althea Arnaquq-Baril's studio's website is at www.unikkaat.com. Ilisaqsivik Society's website is ilisaqsivik.ca. Despite their recognized success and multiple awards, no government or organization has provided them with long-term funding, so they have recently opened a hotel that they hope will provide a stable source of income as well as make it feasible for more outsiders to come to Clyde River. See, e.g., Steve Ducharme, "New Nunavut hotel will funnel profits into long-term charity, *Nunatsiaq News*, August 28, 2015. On Clyde River's legal challenge, see, e.g., *Clyde River (Hamlet) v. Petroleum Geo-Services Inc.*, 2017 SCC 40, https://scc-csc.lexum.com/scc-csc/scc-csc/en/item/16743/index.do; Elyse Skura, "'We Thought No One Would Care': Clyde River Inuit Flooded with Support," CBC News Online, November 29, 2016; Derek Leahy,

"Supreme Court Case Could Change Natural Resource Development in Canada," *National Observer*, December 5, 2016. Inuit Tapiriit Kanatami's website is www.itk.ca and their *National Inuit Suicide Prevention Strategy*, which explicitly calls for social equity as well as "approaches that connect Inuit with our land, culture, and language to foster healing" is at https://www.itk.ca/wp-content/uploads/2016/07/ITK-National-Inuit-Suicide-Prevention-Strategy-2016.pdf. See also the many thoughtful public reflections and calls to action of ITK's current president, Natan Obed. Feeding My Family's Facebook page is at https://www.facebook.com/groups/239422122837039. For an interview with its founder Leesee Papatsie, see "Feeding My Family, an Interview with Leesee Papatsie," *Northern Public Affairs*, Fall 2012, http://www.northernpublicaffairs.ca/index/magazine/archives-volume-1-issue-2/nutrition-north-feeding-my-family-an-interview-with-leesee-papatsie-2/. Tanya Tagaq's most recent album is *Retribution* (2016); for her Twitter feed see @tagaq; for an interview with her on CBC Radio's *Unreserved* program, see "Tanya Tagaq Seeks Retribution with New Album," September 25, 2016, http://www.cbc.ca/radio/unreserved/creativity-that-challenges-the-status-quo-as-it-changes-the-landscape-1.3774044/tanya-tagaq-seeks-retribution-with-new-album-1.3775135.

4. These are two of the eight core Inuit societal values identified by the Government of Nunavut, also called Inuit Piqujangit or Inuit communal laws. See http://www.gov.nu.ca/information/inuit-societal-values.

5. On the critique of Indigenous knowledge as purely local knowledge, see Cameron, "Securing Indigenous Politics," 105–6.

6. Jamesie Mike, interview, August 29, 2008, AVC. Unless otherwise stated, all further references to Jamesie's story of his father are from this interview.

7. For an in-process atlas of Inuit trails, see Aporta et al., *Pan Inuit Trails Atlas*.

8. Kilabuk was the daughter of the whaling leader Veevee and his wife Imaqi. Other Inuit also relocated around the eastern Arctic on whaling ships. See Harper, "The Collaboration of James Mutch and Franz Boas," 64; Hantzsch, *My Life*, 38; Eber, *When the Whalers*; Tyson, *Cruise*.

9. Akpalialuk was Kanajuq's son from his first marriage in Salliq. He had several children but only Akpalialuk came with him. Kanajuq was widowed twice, and Jamesie was one of his sons from his third wife, Atchina. Jamesie and Akpalialuk were therefore half-brothers. Jamesie didn't date the trip to Scotland in calendar years. He said it happened before his father married his mother (Atchina), when families were dispersing from the whaling stations into smaller communities. Kenn Harper gives the 1919 date in "The Albert."

10. Qikiqtani Inuit Association, *Qimmiliriniq*. While many Inuit saw their move into the settlements as temporary and potentially reversible, the loss of dog teams made it more difficult to return to the land, since they then needed to be able to afford snowmobiles and gas. Qikiqtani Inuit Association, *Nuutauniq*, 36–37.

11. Womack, *Red on Red*, 42, as cited in Martin, *Stories in a New Skin*, 6. Original wording says "Indian people," but I have edited as I think it also applies to Inuit and other Indigenous people.

12. M'Donald, *Narrative*, 13–14; Harbsmeier, "Bodies and Voices," 59–60 (mosquitoes); Harper, *Give Me My Father's Body*, 34, 170; Gilberg, "Uisâkavsak," 85–86; Eber, *When the Whalers*, 53; Hendrik, *Memoirs*, 74–81 (house lights quote on 76). For an Inuit perspective on southern Canada in the twentieth century, see Freeman, *Life among the Qallunaat*.

13. Hall, *Arctic Researches*, 102–3.

14. On hardship in the United States, see among many other works Rosenow, *Death and Dying*; Litwack, *Trouble in Mind*; Williams, *They Left Great Marks on Me*; Leflouria, *Chained in Silence*; Blackhawk, *Violence over the Land*; Jacoby, *Shadows at Dawn*; Andrews, *Killing for Coal*; Limerick, *Legacy of Conquest*.

15. For critical discussions of resilience, see, e.g., Crate and Nuttall, eds., *Anthropology and Climate Change*; Marino, "Environmental Migration" and "The Long History"; Cameron, Mearns, and McGrath, "Translating Climate Change"; Hammond, *This Place Will Become Home*, 212–13. On the need to think about "who can stay at home," see Ahmed, Castada, Fortier, and Sheller, *Uprootings/Regroundings*, 7.

Appendix

1. For critiques of Qallunaat research methods, and suggestions and insights, see McGrath, "Isumaksaqsiurutigijakka"; Flaherty, "Freedom of Expression"; de Leeuw, Cameron, and Greenwood, "Participatory and Community-Based Research"; Cameron, *Far Off Metal River*. On research ethics, see also Nagy, ed., "Intellectual Property and Ethics."

2. Kappianaq in *Travelling and Surviving*, 71. On the importance of Inuktitut, see also Dorais, *Language in Inuit Society*, esp. 95.

3. On interpreters in healthcare, see Kaufert and Putsch, "Communication through Interpreters"; Kaufert et al., "Experience of Aboriginal Health Interpreters"; Kaufert et al., "Culture and Informed Consent." On working with translated sources, see Nagy, "Time, Space, and Memory," in Stern and Stevenson, eds., *Critical Inuit Studies*.

4. Andrew reviewed this section and agreed to me including it, although he felt that it made him sound like he knows more than he actually does. I disagree.

5. For an excellent overview of early Qallunaat place names in Cumberland Sound, see Goldring, "Whaling-Era Toponymy."

Works Cited

Archival Sources

Angmarlik Visitor Centre, Pangnirtung, NU [AVC]
    Nakashuk (née Karpik), Margaret. Interviews with Etooangat Aksayuk and Pauloosie Angmarlik, 1994. Interviews are in Inuktitut. I worked with recordings of English translations (in my possession) by Andrew Dialla.
    Routledge, Karen (researcher), and Andrew Dialla (interpreter). Collection of interviews from 2008 with Daisy Dialla, Elisapee Ishulutaq, Peteroosie Karpik, Jamesie Mike, Inuusiq Nashalik, Pauloosie Veevee, and Rosee Veevee.
British Museum [BM]
    Reverend Edgar W. T. Greenshield Photographs.
Canadian Museum of History, Gatineau, QC [CMH]
    Eber, Dorothy. "Inuit Memories of the Whaling Days: Interviews on South Baffin Island." Ms. IV-C-138M.
Explorers' Club Archives and Manuscript Collections, New York, NY [EC]
    Collection of the Lady Franklin Bay Expedition. Accession #2003-007.
General Synod Archives, Anglican Church of Canada [ACC]
    Florence Hirst Journals.
G. W. Blunt Library, Mystic Seaport Museum, Mystic, CT [MSM]
    Buddington Family Collection. Coll. 257.
    Calabretta, Fred. Interview with Jean Bullard, April 27, 1992. OH 92-4.
    Calabretta, Fred. Interview with George A. Comer, March 13, 1984. OH 84-9.
    Ebierbing, Sylvia Grinnell, and her family, Groton, CT, newspaper clippings. MR 58.
    Henry Grinnell Letters. Coll. 8.
    New London Crew Lists: 1803-78.
Houghton Library, Harvard University, Cambridge, MA [HL]
    Daniel Fielding Logbook Collection.
Indian and Colonial Research Center, Old Mystic, CT [ICRC]
    Butler, Eva. "Miscellaneous typed working notes by Mrs. Butler regarding Hannah and Joe Ebierbing." Part of Eva Butler's notebook collection.
    Colby, Barnard L. "Tells Intimate Stories about Groton Eskimos," clipping, DOC 1939-03-00A.

Colby, Barnard L. "Joe and Hannah, Famous Eskimos, Lived in Groton," clipping, DOC 1939-04-00A.
Joseph Ebierbing Collection. Ms Eb47 and MsBd Eb47.
Library of Congress, Washington, DC [LOC]
    A. W. Greely Papers.
    Documentary Photographs of the U.S. Expedition to Lady Franklin Bay, Grinnell Land, 1881–84. Lot 4262.
National Archives and Records Administration, College Park, MD [NARA MD]
    Brainard Collection of the Lady Franklin Bay Expedition (200.13-LFB).
    George E. Tyson Papers.
    Records of the Lady Franklin Bay Expedition (RG-27.4.5).
National Archives and Records Administration, Waltham, MA [NARA MA]
    Crew lists for New Bedford, MA, 1820–1939 (RG36).
    Crew lists for New London, CT, 1838–85 (RG36).
National Archives of Canada, Ottawa, ON [LAC]
    Church Missionary Society fonds. MG17-B2.
    National Film Board fonds. RG53.
    Records of the Northwest Territories and Yukon Branch. RG85-C-1-a.
    Records of the Royal Canadian Mounted Police. RG18.
    T. H. Tredgold fonds. MG30-E-632.
Hudson's Bay Company Archives, Winnipeg, MB [HBCA]
    Pangnirtung Post Journals.
Archives Center, National Museum of American History, Smithsonian Institution, Washington, DC [NMAH]
    Charles Francis Hall Collection.
New Bedford Whaling Museum, New Bedford, MA [NBWM]
    Accounts of the *Black Eagle,* 1860s (Mss 79).
    Accounts of the *Helen F.* and *Nile,* 1874–76 (IMA 795).
    Ship logbooks and journals:
        Andrews 1867.
        Ansel Gibbs 1860–61.
        Antelope 1861–63.
        Antelope 1865–66.
        Black Eagle 1860–61.
        Daniel Webster 1860–63.
        Era 1891–92.
        Franklin 1878–79.
        George Henry 1860–61.
        Isabella 1867–68.
        Mattapoisett 1878–79.
        Milwood 1867–68.
        S. B. Howes 1873.
New London County Historical Society, New London, CT [NLCHS]
    Calabretta, Fred. "Captain George Comer and the Arctic," B C734.
    R. B. Wall scrapbooks of newspaper clippings.
Northwest Territories Archives, Yellowknife, NT [NTA]

Northwest Territories Department of Justice and Public Services fonds. Jaypeetee Akpalialuk and Marc Stevenson interviews. G-1985-007.

Uvdloriaq, Inuuterssuaq. "The Narrative of Qitdlarssuaq." Translated by Kenn and Navarana Harper. In Robin McGrath Research Collection. N-1998-047: 1-2.

William Wakeham fonds.

Providence Public Library, Providence, RI [PPL]

    Nicholson Whaling Collection:

        Ambrose Bates's journal on the *U.D.*, 1868-69 [also contains entries from his 1867-68 *Milwood* voyage].

        Andrew D. West's journal on the *A. R. Tucker*, 1891-92.

Rauner Special Collections, Dartmouth College, Hanover, NH [RSC]

    David L. Brainard Papers.

Rigsarkivet (Danish National Archives), Copenhagen, Denmark [RA]

    Dansk Demografisk Database. www.ddd.dda.dk/

    Direktorat Kgl. Grønlandske Handel, Bogholder- og Korrespondancekontor, 1820-1901: Dagbøger af kolonibestyrere, assistenter m. fl. i Grønland 1774-1921.

    Ekstrakter af breve fra Nordgrønland med resolutioner 1797-1910.

    Kopier av Nordgronlands inspektorats korrespondance 1871-79 and 1879-88.

University of Alberta Archives, Edmonton, AB [UAA]

    J. Dewey Soper fonds.

Published Sources

Adams, Bluford. *E Pluribus Barnum: The Great Showman and the Making of US Popular Culture*. Minneapolis: University of Minnesota Press, 1997.

Ahmed, Sara. "Home and Away: Narratives of Migration and Estrangement." *International Journal of Culture Studies* 2:3 (1999): 329-47.

Ahmed, Sara, Claudia Castada, Anne-Marie Fortier, and Mimi Sheller. *Uprootings/Regroundings: Questions of Home and Migration*. New York: Berg, 2003.

Alia, Valerie. *Names and Nunavut: Culture and Identity in the Inuit Homeland*. New York: Berghahn Books, 2007.

Alix, Claire, ed. "Arctic Peoples and Wood." Special issue, *Études Inuit Studies* 36:1 (2012).

Anawak, Jack. "The Environment." Canada. Parliament. House of Commons. Ed. Hansard. 34th Parliament, 3rd Session. Vol. 8. May 7, 1992.

Anbinder, Tyler. *Five Points: The Nineteenth-Century New York City Neighborhood That Invented Tap Dance, Stole Elections, and Became the World's Most Notorious Slum*. New York: Free Press, 2001.

Andrews, Thomas G. *Killing for Coal: America's Deadliest Labor War*. Cambridge, MA: Harvard University Press, 2010.

Aporta, Claudio. "Routes, Trails, and Tracks: Trail-Breaking among the Inuit of Igloolik." *Études/Inuit/Studies* 28:2 (2004): 9-38.

———. "The Trail as Home: Inuit and Their Pan-Arctic Network of Routes." *Human Ecology* 37 (2009): 131-46.

Aporta, Claudio, Michael Bravo, and Fraser Taylor. *Pan Inuit Trails Atlas*. www.paninuittrails.org.

Arnaktauyok, Germaine, ed. *Stories from Pangnirtung*. Edmonton: Hurtig, 1976.

Aupilaarjuk, Mariano, et al. *Inuit Qaujimajatuqangit: Shamanism and Reintegrating Wrongdoers into the Community.* Ed. Jarich Oosten and Frédéric Laugrand. Iqaluit: Nunavut Arctic College, 2002.

Aupilaarjuk, Mariano, and Tulimaaq, et al. *Cosmology and Shamanism.* Ed. Bernard Saladin d'Anglure. Iqaluit: Nunavut Arctic College, 2001.

Bachelard, Gaston. *La terre et les rêveries de la volonté.* Paris: J. Corti, 1948.

Barnum, Phineas T. *An Illustrated Catalogue and Guide Book to Barnum's American Museum.* New York: Wynkoop, Hallenbeck and Thomas, [186–?]. http://www.disabilitymuseum.org /lib/docs/872.htm.

Barr, William. *The Expeditions of the First International Polar Year, 1882–1883.* Calgary: Arctic Institute of North America, 2008.

———. "The Eighteenth-Century Trade between the Ships of the Hudson's Bay Company and the Hudson Strait Inuit." *Arctic* 47:3 (1994): 236–46.

Barron, William. *Old Whaling Days.* Hull: William Andrews, 1895.

Baudrillard, Jean. *America.* London: Verso, 1989.

Bennett, John, and Susan Rowley, eds. *Uqalurait: An Oral History of Nunavut.* Montreal: McGill-Queen's University Press, 2004.

Bennett, John. "Whalers, Missionaries, and Inuit in Cumberland Sound." Master's thesis, Carleton University, 1985.

Bek, Adam. "Akilinermut avalagkaluartup nunaminut uterame okalungusiutai." *Atuagagdliutit.* Serialized in 5 (61) through 6 (64): 966–1010. Partial draft translation courtesy of Kenn Harper.

Belyea, Barbara. "Captain Franklin in Search of the Picturesque." *Essays on Canadian Writing* 40 (Spring 1990): 1–24.

Berger, Thomas. *Northern Frontier, Northern Homeland: The Report of the Mackenzie Valley Pipeline Inquiry.* Vol. 1. Ottawa: Minister of Supply and Services Canada, 1977.

Berman, Marshall. *All That Is Solid Melts into Air: The Experience of Modernity.* New York: Penguin, 1988.

Bernier, J. E. *Report on the Dominion Government Expedition to the Arctic Islands and Hudson Strait on Board the C.G.S. Arctic 1906–1907.* Ottawa: C. H. Parmalee, 1909.

Bessels, Emil. "Smith Sound, and Its Exploration." *Proceedings of the United States Naval Institute* 10:3 (1884): 333–447.

Blackhawk, Ned. *Violence over the Land: Indians and Empires in the Early American West.* Cambridge, MA: Harvard University Press, 2009.

Black-Rogers, Mary. "'Varieties of Starving': Semantics and Survival in the Subarctic Fur Trade, 1750–1850." *Ethnohistory* 33 (1986): 353–83.

Blake, E. Vale, ed. *Arctic Experiences: Containing Capt. George E. Tyson's Wonderful Drift on the Ice-Floe, a History of the Polaris Expedition, the Cruise of the Tigress, and Rescue of the Polaris Survivors. To Which Is Added a General Arctic Chronology.* New York: Harper and Brothers, 1874.

Bloom, Lisa. *Gender on Ice: American Ideologies of Polar Expeditions.* Minneapolis: University of Minnesota Press, 1993.

Bocking, Stephen, and Brad Martin. *Ice Blink: Navigating Northern Environmental History.* Calgary: University of Calgary Press, 2016.

Blunt, Alison, and Robyn Dowling. *Home.* London: Routledge, 2006.

Boas, Franz. *Arctic Expedition, 1883–1884: Translated German Newspaper Accounts of My Life with the Eskimos.* Ed. Norman F. Boas and Doris W. Boas. Mystic, CT: [n.p.], 2009.

———. *The Central Eskimo.* Lincoln: University of Nebraska Press, 1964.

———. "The Eskimo of Baffin Land and Hudson Bay." *Bulletin of the American Museum of Natural History* 15 (1901): 276–79.

Bockstoce, John R. *Whales, Ice, and Men: The History of Whaling in the Western Arctic.* Seattle: University of Washington Press, 1986.

Bogdan, Robert. *Freak Show: Presenting Human Oddities for Amusement and Profit.* Chicago: University of Chicago Press, 1988.

Bordin, Guy. "La nuit inuit. Elements de réflexion." *Etudes/Inuit/Studies* 26:1 (2002): 45–70.

Bown, Stephen R. *Scurvy: How a Surgeon, a Mariner, and a Gentleman Solved the Greatest Medical Mystery of the Age of Sail.* New York: St. Martin's Press, 2003.

Boydston, Jeanne. *Home and Work: Housework, Wages, and the Ideology of Labor in the Early Republic.* New York: Oxford University Press, 1990.

Bradbury, Bettina. "The Fragmented Family: Family Strategies in the Face of Death, Illness, and Poverty, Montreal, 1860–1885." In *Childhood and Family in Canadian History*, ed. Joy Parr. Toronto: McLelland and Stewart, 1982: 109–28.

Brainard, David L., and Bessie Rowland James. *Six Came Back: The Arctic Adventure of David L. Brainard.* Indianapolis: Bobbs-Merrill, 1940.

Bravo, Michael. "Sea Ice Mapping: Ontology, Mechanics, and Human Rights at the Ice Floe Edge." In *High Places: Cultural Geographies of Mountains and Ice*, 161–76. Ed. D. Cosgrove and V. della Dora. London: IB Tauris, 2008.

Brickell, Katherine. "'Mapping' and 'Doing' Critical Geographies of Home." *Progress in Human Geography* 36:2 (2012): 225–44.

Briganti, Chiara, and Kathy Mezei, eds. *The Domestic Space Reader.* Toronto: University of Toronto, 2012.

Briggs, Jean. *Never in Anger: Portrait of an Eskimo Family.* Cambridge, MA: Harvard University Press, 1970.

Brody, Hugh. *The Other Side of Eden: Hunter-Gatherers, Farmers, and the Shaping of the World.* London: Faber and Faber, 2001.

———. *The People's Land: Eskimos and Whites in the Eastern Arctic.* Markham, ON: Penguin Books, 1975.

Burrows, Edwin G., and Mike Wallace. *Gotham: A History of New York City to 1898.* New York: Oxford University Press, 1999.

Burt, Page M. *Barrenland Beauties: Showy Plants of the Canadian Arctic.* Yellowknife, NT: Outcrop, 2001.

Busch, Briton Cooper. *Whaling Will Never Do For Me: The American Whaleman in the Nineteenth Century.* Lexington: University Press of Kentucky, 2014 (1994).

Cameron, Emilie. *Far Off Metal River: Inuit Lands, Settler Stories, and the Making of the Contemporary Arctic.* Vancouver: UBC Press, 2016.

———. "Securing Indigenous Politics: A Critique of the Vulnerability and Adaptation Approach to the Human Dimensions of Climate Change in the Canadian Arctic." *Global Environmental Change* 22:1 (2012): 103–14.

Cameron, Emilie, Rebecca Mearns, and Janet Tamalik McGrath. "Translating Climate Change: Adaptation, Resilience, and Climate Politics in Nunavut, Canada." *Annals of the Association of American Geographers* 105:2 (2015): 274–83.

Campbell, Robert. *In Darkest Alaska: Travel and Empire Along the Inside Passage.* Philadelphia: University of Pennsylvania, 2007.

Canada. Welfare Division. Northern Administration Branch. *The Q Book.* Ottawa: Ministry of Northern Affairs and Natural Resources, Government of Canada, 1964.

Cardno, David Hawthorn. *A Whaler's Tale: The Memoirs of David Hawthorn Cardno of Peterhead, 1853–1938*. Ed. Gavin Sutherland. Mintlaw, Scotland: Aberdeenshire Council, 1996.

Carpenter, Kenneth J. *The History of Scurvy and Vitamin C*. New York: Cambridge University Press, 1986.

Carlson, Hans M. *Home Is the Hunter: The James Bay Cree and Their Land*. Vancouver: UBC Press, 2008.

Cavell, Janice. "Publishing Sir John Franklin's Fate: Cannibalism, Journalism, and the 1881 Edition of Leopold McClintock's *The Voyage of the "Fox" in the Arctic Seas*. In *Book History* 16 (2013): 155–84.

———. "Going Native in the North: Reconsidering British Attitudes during the Franklin Search, 1848–1859." *Polar Record* 45:232 (2009): 25–35.

———. *Tracing the Connected Narrative: Arctic Exploration in British Print Culture, 1818–1860*. Toronto: University of Toronto Press, 2008.

Chapin, Robert Coit. *The Standard of Living among Workingmen's Families in New York City*. Arno Press and the *New York Times*: New York, 1971 (1909).

Clare, Stephanie. "Feeling Cold: Phenomenology, Spatiality, and the Politics of Sensation." *Differences: A Journal of Feminist Cultural Studies*. 24:1 (2013): 169–91.

Clark, Captain G. V. *The Last of the Whaling Captains*. Glasgow: Brown, Son and Ferguson, 1986.

Clayton, Daniel. *Islands of Truth: The Imperial Fashioning of Vancouver Island*. Vancouver: UBC Press, 2000.

Clendinnen, Inga. *Ambivalent Conquests: Maya and Spaniard in Yucatan, 1517–1570*. New York: Cambridge University Press, 1987.

Coates, Kenneth S., and William R. Morrison. "The New North in Canadian History and Historiography." *History Compass* 6 (2008): 639–58.

Coates, Peter. *American Perceptions of Immigrant and Invasive Species: Strangers on the Land*. Berkeley: University of California Press, 2006.

Colby, Barnard L. *For Oil and Buggy Whips: Whaling Captains of New London County, Connecticut*. Mystic, CT: Mystic Seaport Museum, 1990.

Cole, Douglas, and Ludger Müller-Wille. "Franz Boas' Expedition to Baffin Island, 1883–1884." *Etudes/Inuit/Studies* 8:1 (1984): 37–63.

Collignon, Béatrice. *Knowing Places: The Inuinnait, Landscapes and the Environment*. Translated by Linna Weber Müller-Wille. [Edmonton, AB]: CCI Press, 2006.

Comer, George, and W. Gillies Ross. *An Arctic Whaling Diary: The Journal of Captain George Comer in Hudson Bay, 1903–1905*. Toronto: University of Toronto Press, 1984.

Cook, James W. *The Arts of Deception: Playing with Fraud in the Age of Barnum*. Cambridge, MA: Harvard University Press, 2001.

———. "Of Men, Missing Links, and Nondescripts: The Strange Career of P. T. Barnum's "What Is It?" Exhibition. In *Freakery: Cultural Spectacles of the Extraordinary Body*. Edited by Rosemary Garland Thomson. New York: New York University Press, 1996.

Craciun, Adriana. *Writing Arctic Disaster: Authorship and Exploration*. Cambridge: Cambridge University Press, 2016.

Crate, Susan Alexandra, and Mark Nuttall, eds. *Anthropology and Climate Change: From Encounters to Actions*. Walnut Creek, CA: Left Coast Press, 2009.

Creighton, Margaret S. *Rites and Passages: The Experience of American Whaling, 1830–1870*. New York: Cambridge University Press, 1997.

Cronon, William, ed. *Uncommon Ground: Rethinking the Human Place in Nature*. New York: Norton, 1996.

———. *Nature's Metropolis: Chicago and the Great West*. New York: Norton, 1991.
Cruikshank, Julie. *Do Glaciers Listen? Local Knowledge, Colonial Encounters and Social Imagination*. Vancouver: UBC Press, 2005.
———. "Images of Society in Klondike Gold Rush Narratives: Skookum Jim and the Discovery of Gold." *Ethnohistory* 39:1 (1992): 20–41.
Cull, Richard. "A Description of Three Esquimaux from Kinnooksook, Hogarth's Sound, Cumberland Strait." *Journal of the Ethnological Society of London* 4 (1856): 215–25.
Daniel, Thomas M. *Captain of Death: The Story of Tuberculosis*. Rochester, NY: University of Rochester Press, 2006.
Daschuk, James. *Clearing the Plains: Disease, Politics of Starvation, and the Loss of Aboriginal Life*. Regina: University of Regina Press, 2013.
David, Robert G. *The Arctic in the British Imagination, 1818–1914*. Manchester, UK: Manchester University Press, 2000.
Davidson, Joyce, Liz Bondi, and Mick Smith, eds. *Emotional Geographies*. Burlington, VT: Ashgate, 2007.
Davis, C. H., and Charles Francis Hall. *Narrative of the North Polar Expedition*. Washington, DC: Government Printing Office, 1876.
Davis, Lance E., et al., *In Pursuit of Leviathan: Technology, Institutions, Productivity, and Profits in American Whaling, 1816–1906*. Chicago: University of Chicago Press, 1997.
Davis-Fisch, Heather. *Loss and Cultural Remains in Performance: The Ghosts of the Franklin Expedition*. New York: Palgrave Macmillan, 2012.
de Leeuw, Sarah, Emilie S. Cameron, and Margo L. Greenwood. "Participatory and Community-Based Research, Indigenous Geographies, and the Spaces of Friendship: A Critical Engagement." *Canadian Geographer* 56 (2012): 180–94.
Decker, Robert. *The Whaling City: A History of New London*. Chester, CT: Published for the New London Historical Society by Pequot Press, 1976.
Dennett, Andrea. *Weird and Wonderful: The Dime Museum in America*. New York: New York University Press, 1997.
Dick, Lyle. *Muskox Land: Ellesmere Island in the Age of Contact*. Calgary: University of Calgary Press, 2001.
———. "The Fort Conger Shelters and Vernacular Adaptation to the High Arctic." *Society for the Study of Architecture in Canada (SSAC) Bulletin* 16:1 (1991): 13–23.
Dodge, Ernest S., and C. C. Loomis. "Hall, Charles Francis." *Dictionary of Canadian Biography Online*. Toronto: University of Toronto/Université Laval, 2000.
Dolin, Eric Jay. *Leviathan: The History of Whaling in America*. New York: Norton, 2007.
Dorais, Louis-Jacques. *Language and Inuit Society*. Iqaluit: Nunavut Arctic College, 1996.
Dunlap, Thomas R. *Nature and the English Diaspora: Environment and History in the United States, Canada, Australia, and New Zealand*. Cambridge: Cambridge University Press, 1999.
Eber, Dorothy. *When the Whalers Were Up North: Inuit Memories from the Eastern Arctic*. Montreal: McGill-Queen's University Press, 1989.
Edwards, Laura F. *Gendered Strife and Confusion: The Political Culture of Reconstruction*. Champaign, IL: University of Illinois Press, 1998.
Ekho, Naqi, and Uqsuralik Ottokie. *Childrearing Practices*. Ed. Jean Briggs. Iqaluit: Nunavut Arctic College, 2000.
Ellis, M. A. B. "Presence and Absence: An Exploration of Scurvy in the Commingled Subadults in the Spring Street Presbyterian Church Collection, Lower Manhattan." In *International Journal of Osteoarchaeology* 26 (2016): 759–66.

Emerson, Haven. "Sunlight and Health." *American Journal of Public Health* (1933): 437–40.
Erickson, Kai. "Preface." In *Animal Disease and Human Trauma: Emotional Geographies of Disaster*. Ed. Ian Convery, Maggie Mort, Josephine Baxter, and Cathy Bailey. New York: Palgrave Macmillan, 2008: vii–xvii.
Eustace, Nicole, et al. "AHR Conversation: The Historical Study of Emotions." *American Historical Review* 117 (2012): 1487–1531.
Faulkner, Joseph. *Eighteen Months on a Greenland Whaler*. New York: [private publication], 1878.
Feeney, Robert Earl. *Polar Journeys: The Role of Food and Nutrition in Early Exploration*. Fairbanks: University of Alaska Press, 1997.
Fienup-Riordan, Ann. *Eskimo Essays: Yup'ik Lives and How We See Them*. New Brunswick, NJ: Rutgers University Press, 1990.
Finley, K. J. "Natural History and Conservation of the Greenland Whale, or Bowhead, in the Northwest Atlantic." *Arctic* 54 (2001): 55–76.
Flaherty, Martha. "Freedom of Expression or Freedom of Exploitation?" *Northern Review* 14 (Summer 1995): 178–85.
Fleming, Archibald Lang. *Perils of the Polar Pack*. Toronto: Missionary Society of the Church of England, 1932.
Flora, Janne. "The Lonely Un-Dead and Returning Suicide in Northwest Greenland." In *Suicide and Agency: Anthropological Perspectives on Self-Destruction, Personhood, and Power*. Ed. Ludek Broz and Daniel Münster. London: Ashgate, 2015.
Freeman, Minnie Aodla. *Life among the Qallunaat*. Ed. Keavy Martin, Julie Rak, and Norma Dunning. Winnipeg: University of Manitoba Press, 2015.
Friedland, Roger, and Deirdre Boden. *NowHere: Space, Time, and Modernity*. Berkeley: University of California Press, 1994.
Friesen, T. Max, and Charles D. Arnold. "The Timing of the Thule Migration: New Dates from the Western Arctic." *American Antiquity* 73:3 (July 2008): 527–38.
Garrison, Laurie. "Virtual Reality and the Subjective Response: Narrating the Search for the Franklin Expedition through Robert Burford's Panorama." *Early Popular Visual Culture* 10 (2012): 7–22.
Gearheard, Shari Fox, et al., eds. *The Meaning of Ice: People and Sea Ice in Three Arctic Communities*. Montreal: International Polar Institute, 2013.
Geraci, Joseph R., and Thomas G. Smith. "Vitamin C in the Diet of Inuit Hunters from Holman, Northwest Territories." *Arctic* 32 (1979): 135–39.
Gilder, William Henry. "Dangers of the Ice Pack." *Cosmopolitan* 4:4(1887): 276–81.
———. *Schwatka's Search: Sledging in the Arctic in Quest of the Franklin Records*. New York: C. Scribner's Sons, 1881.
Gilberg, Rolf. "Uisâkavsak, 'The Big Liar.'" *Folk* 11/12 (1970): 83–95.
Glennie, Paul, and Nigel Thrift. "Time-Geography." In *International Encyclopedia of Social and Behavioral Sciences*. Ed. Paul B. Baltes and James D. Wright. Amsterdam: Elsevier, 2001: 15692–96.
———. "Re-working E. P. Thompson's 'Time, Work-Discipline, and Industrial Capitalism.'" *Time and Society* 5:3(1996): 275–99.
Goddard, Ives. "Synonymy." In *Handbook of North American Indians*. Vol. 5: *Arctic*. Ed. David Damas. Washington, DC: Smithsonian Institution, 1984, 5–7.
Goldring, Philip. "Historians and Inuit: Learning from the Qikiqtani Truth Commission, 2007–2010." *Canadian Journal of History* 50:3 (2015): 492–523.

———. "Southeast Baffin Island Historical Reports." Ottawa, Canadian Parks Service, unpublished manuscript, 1988.

———."Inuit Economic Responses to Euro-American Contacts: Southeast Baffin Island, 1824–1940." *Journal of the Canadian Historical Association: Historical Papers* 21:1 (1986): 146–72.

———. "The Last Voyage of the McLellan." *Beaver* January/February 1986: 39–44.

———. "Whaling-Era Toponymy of Cumberland Sound," *Canoma* 11:2 (1985): 28–34.

Grace, Sherrill E. *Canada and the Idea of North*. Montreal: McGill-Queen's University Press, 2002.

Grant, Shelagh D. *Polar Imperative: A History of Arctic Sovereignty in North America*. Toronto: Douglas and McIntyre, 2010.

———. *Arctic Justice: On Trial for Murder, Pond Inlet, 1923*. Montreal: McGill-Queen's University Press, 2002.

Greely, Adolphus W. Report of the Proceedings of the United States Expedition to Lady Franklin Bay, Grinnell Land. Washington: Government Printing Office, 1888.

———. *Three Years of Arctic Service: An Account of the Lady Franklin Bay Expedition of 1881–84 and the Attainment of the Farthest North*. 2 vols. New York: C. Scribner's, 1886.

Greenblatt, Stephen. *Marvelous Possessions: The Wonder of the New World*. Chicago: University of Chicago Press, 1991.

Green, Edward H. "Medical Report on the Condition of the Survivors of the Greely Party, When Rescued by the Relief Squadron." *Medical News* 45:10 (1884): 260–64.

Gregory, Derek. "Imaginative Geographies." In *The Dictionary of Human Geography*. Ed. Derek Gregory et al. New York: Wiley-Blackwell, 2009.

Grønnow, Bjarne. "Blessings and Horrors of the Interior: Ethno-Historical Studies of Inuit Perceptions Concerning the Inland Region of West Greenland." *Arctic Anthropology* 46:1–2 (2009): 191–201.

Grygier, Pat Sandiford. *A Long Way from Home: The Tuberculosis Epidemic among the Inuit*. Toronto: McGill-Queen's University Press, 1994.

Guttridge, Leonard F. *Ghosts of Cape Sabine: The Harrowing True Story of the Greely Expedition*. New York: Putnam, 2000.

Hackett, Paul. *A Very Remarkable Sickness: Epidemics in the Petit Nord, 1670 to 1846*. Winnipeg: University of Manitoba Press, 2002.

Hall, Charles Francis. *Arctic Researches, and Life among the Esquimaux: Being the Narrative of an Expedition in Search of Sir John Franklin, in the Years 1860, 1861, 1862*. New York: Harper and Brothers, 1865.

Hammond, Laura. *This Place Will Become Home: Refugee Repatriation to Ethiopia*. Ithaca, NY: Cornell University Press, 2004.

Hantzsch, Bernhard Adolph. *My Life Among the Eskimos: Baffinland Journeys in the Years 1909 to 1911*. Ed. Leslie H. Neatby. Saskatoon: University of Saskatchewan, 1977.

Harbsmeier, Michael. "Bodies and Voices from Ultima Thule: Inuit Explorations of the Kablunat from Christian IV to Knud Rasmussen." In *Narrating the Arctic: A Cultural History of Nordic Scientific Practices*. Ed. Michael Bravo and Sverker Sörlin. Canton, MA: Science History Publications, 2002: 33–71.

Hareven, Tamara. "The Home and Family in Historical Perspective." *Social Research* (1991): 253–85.

———. *Family Time and Industrial Time: The Relationship between Family and Work in a New England Industrial Community*. Cambridge: Cambridge University Press, 1982.

Harris, R. Cole, and Elizabeth Phillips, eds. *Letters from Windermere 1912–1914*. Vancouver: UBC Press, 68.

Harper, Kenn. *In Those Days: Collected Writings on Arctic History*. Iqaluit: Inhabit Media, 2013.

———. "The Collaboration of James Mutch and Franz Boas, 1883–1922." *Etudes/Inuit/Studies* 32:2 (2008): 53–71.

———. "March 18, 1875—The Death of a Daughter." *Nunatsiaq News*, March 17, 2006.

———. "Uquuquq a Reluctant Traveller: A Contribution to Inuit History." *Above and Beyond* (March/April 2005).

———. *Give Me My Father's Body: The Life of Minik, the New York Eskimo*. South Royalton, VT: Steerforth Press, 2000.

———. "The *Albert*—A Famous Arctic Ship." *Nunatsiaq News*, May 27, 1995.

———. "William Duval (1858–1931). In *Arctic* 38:1 (1985): 74–75.

———. "The Burning of the Easonian." *North/Nord* (May/June 1974).

———. *Pangnirtung*. Arctic Bay, NU: s.n., 1972.

Harvey, David. "Between Space and Time: Reflections on the Geographical Imagination." *Annals of the Association of American Geographers* 80:3 (1990): 418–34.

Hatton, Joseph, and Moses Harvey. Newfoundland: Its History, Its Present Condition, and Its Prospects for the Future. Boston: Cupples and Hurd, 1888.

Hastrup, Kirsten. "Hunger and the Hardness of Facts." *Man* 28 (1993): 727–39.

Hasty, William, and Kimberley Peters. "The Ship in Geography and Geographies of Ships." *Geography Compass* 6:11 (2012): 660–76.

Hayes, Isaac. The Open Polar Sea: A Narrative of the Voyage of Discovery towards the North Pole. London: Sampson Low, Son, and Marston, 1867.

Hays, J. N. *Epidemics and Pandemics: Their Impacts on Human History*. Santa Barbara: ABC-CLIO, Inc., 2005.

Heighton, Stephen. *Afterlands*. Toronto: Knopf Canada, 2005.

Henderson, Bruce B. *Fatal North: Adventure and Survival aboard USS Polaris, the First U.S. Expedition to the North Pole*. New York: New American Library, 2001.

Hendrik, Hans. *Memoirs of Hans Hendrik, the Arctic Traveller, Serving under Kane, Hayes, Hall, and Nares, Written by Himself*. Trans. Hinrich Rink. Ed. George Stephens. London: Trübner, 1878.

Henkin, David M. *Postal Age: The Emergence of Modern Communications in Nineteenth-Century America*. Chicago: University of Chicago Press, 2007.

———. *City Reading: Written Words and Public Spaces in Antebellum New York*. New York: Columbia University Press, 1998.

Hill, Jen. *White Horizon: The Arctic in the Nineteenth-Century British Imagination*. Albany: SUNY Press, 2009.

Holland, Peter. *Home in the Howling Wilderness: Settlers and the Environment in Southern New Zealand*. Auckland: Auckland University Press, 2013.

Ignatiev, Noel. *How the Irish Became White*. New York: Routledge, 1995.

Inuit Circumpolar Council—Canada. "The Sea Ice Is Our Highway: An Inuit Perspective on Transportation in the Arctic." Inuit Circumpolar Council, 2008. http://psc.apl.washington.edu/HLD/ArcticChangell/20080423_iccamsa_finalpdfprint.pdf.

Inuit Tapiriit Kanatami. *National Inuit Suicide Prevention Strategy*. [Ottawa]: [Inuit Tapiriit Kanatami], 2016. Available online at www.itk.ca.

———. "Maps of Inuit Nunangat." https://www.itk.ca/publication/maps-inuit-nunangat-inuit-regions-canada.

———. [Formerly Inuit Tapirisat of Canada]. *Nunatsiaq: The Good Land.* Film reel. Ottawa: Inuit Tapirisat of Canada, 1976.

Ipellie, Alootook. *Arctic Dreams and Nightmares.* Penticton, BC: Theytus Books, 1993.

———. "Nunatsiaqmiut: People of the Good Land." Pts. 1, 2. *Inuit Art Quarterly* 7:2 (1992): 14–21 and 7:3 (1992): 22–29.

Irniq, Peter. "Inuit Naming." Blog Channel, IsumaTV. No date. http://www.isuma.tv/our-changing-language/inuit-naming.

Jacobson, Matthew Frye. *Whiteness of a Different Color: European Immigrants and the Alchemy of Race.* Cambridge, MA: Harvard University Press, 1998.

Jacoby, Karl. *Shadows at Dawn: An Apache Massacre and the Violence of History.* New York: Penguin, 2009.

Jensen, Einar Lund, Kristine Raahauge, and Hans Christian Gulløv. *Cultural Encounters at Cape Farewell: The East Greenlandic Immigrants and the German Moravian Mission in the Nineteenth Century.* Copenhagen: Museum Tuscalanum Press, 2011.

Johnson, Susan. *Roaring Camp: The Social World of the California Gold Rush.* New York: W.W. Norton, 2000.

Johnson, Walter. "Time and Revolution in African America: Temporality and the History of Atlantic Slavery." In *Rethinking American History in a Global Age*, ed. Thomas Bender, 148–67. Berkeley: University of California Press, 2002.

Jones, David S. Rationalizing Epidemics: Meanings and Uses of American Indian Mortality since 1600. Cambridge, MA: Harvard University Press, 2004.

———. "Virgin Soils Revisited." *William and Mary Quarterly* 60:4 (2003): 703–42.

Kan, Sergei, ed. *Strangers to Relatives: The Adoption and Naming of Anthropologists in Native North America.* Lincoln: University of Nebraska Press, 2001.

Kane, Elisha Kent. *The U.S. Grinnell Expedition in Search of Sir John Franklin: A Personal Narrative.* New York: Harper and Brothers, 1854.

Kaplan, Susan A., and Robert McCracken Peck. *North by Degree: New Perspectives on Arctic Exploration.* Philadelphia: American Philosophical Society/Lightning Rod Press, 2013.

Kappianaq, George, and Cornelius Nutaraq. *Travelling and Surviving on Our Land.* Ed. Jarich Oosten and Frédéric Legrand. Iqaluit: Nunavut Arctic College, 2001.

Kappianaq, George, Felix Pisuq, and Salome Ka&&ak Qalasiq. *Dreams and Dream Interpretation.* Ed. Stéphane Kolb and Samuel Law. Iqaluit: Nunavut Arctic College, 2001.

Kaplan, Amy. *The Anarchy of Empire in the Making of U.S. Culture.* Cambridge, MA: Harvard University Press, 2005.

Kaufert, Joseph M., and Robert W. Putsch. "Communication through Interpreters in Healthcare: Ethical Dilemmas Arising from Differences in Class, Culture, Language and Power." *Journal of Clinical Ethics* 8, no. 1 (1997): 71–87.

Kaufert, Joseph M., Margaret Lavallée, William W. Koolage, and John O'Neil. "Culture and Informed Consent: The Role of Aboriginal Interpreters in Patient Advocacy in Urban Hospitals." *Issues in the North* 1 (1996): 89–94.

Kaufert, Joseph M., Robert W. Putsch, and Margaret Lavallée. "Experience of Aboriginal Health Interpreters in Mediation of Conflicting Values in End-of-Life Decision Making." *International Journal of Circumpolar Health* 57, Supplement 1 (1998): 43–48.

Keeling, Arn and John Sandlos, eds. *Mining and Communities in Northern Canada: History, Politics, and Memory.* Calgary: University of Calgary Press, 2015.

Keenleyside, Anne. "Euro-American Whaling in the Canadian Arctic: Its Effects on Eskimo Health." *Arctic Anthropology* 27:1 (1990): 1–19.

Kern, Stephen. *The Culture of Time and Space, 1880–1918*. Cambridge, MA: Harvard University Press, 1983.

Kersting, Rudolf, ed. *The White World: Life and Adventures within the Arctic Circle Portrayed by Famous Living Explorers*. New York: Lewis, 1902.

King, J. C. H., and Henrietta Lidchi, eds. *Imaging the Arctic*. Vancouver: UBC Press, 1998.

Kiple, Kenneth F., ed. *The Cambridge Historical Dictionary of Disease*. New York: Cambridge University Press, 2003.

Kleivan, I., and B. Sonne. *Eskimos: Greenland and Canada: Iconography of Religions VIII*. Leiden: E. J. Brill, 1985.

Knötsch, Carol Cathleen. "Franz Boas' Research Trip to Baffin Island, 1882–1884." *Polar Geography and Geology* 17:1 (1993): 3–54.

Kollin, Susan. *Nature's State: Imagining Alaska as the Last Frontier*. Chapel Hill: University of North Carolina Press, 2001.

Krupnik, Igor et al., eds. *SIKU: Knowing Our Ice: Documenting Inuit Sea-Ice Knowledge and Use*. New York: Springer, 2010.

Kublu, Alexina. "Changing Perspectives of Name and Identity among the Inuit of Northeast Canada." In *Arctic Identities: Continuity and Change in Inuit and Saami Societies*, 56–78. Ed. Jarich G. Oosten and Cornelius H. W. Remie. Leiden: Universitet Leiden, 1999.

Kulchyski, Peter, and Frank James Tester, *Kiumajut (Talking Back): Game Management and Inuit Rights, 1900–1970*. Vancouver: UBC Press, 2007.

Kumlien, Ludwig. "Fragmentary Notes on the Eskimo of Cumberland Sound." *Science* 1 (1880): 85–88.

———. *Contributions to the Natural History of Arctic America, Made in Connection with the Howgate Polar Expedition, 1877–78*. Washington, DC: Government Printing Office, 1879.

Kupperman, Karen Ordahl. *Indians and English: Facing off in Early America*. Ithaca, NY: Cornell University Press, 2000.

Kuptana, Rosemarie. "Ilira, or Why it was Unthinkable for Inuit to Challenge Qallunaat Authority." *Inuit Art Quarterly* 8:3 (1993): 5–7.

Lackenbauer, Whitney. *Arctic Front, Arctic Homeland: Re-evaluating Canada's Past Record and Future Prospects in the Circumpolar North*. Toronto: Canadian International Council, 2008.

Laidler, Gita. "Ice through Inuit Eyes: Characterizing the Importance of Sea Ice Processes, Use, and Change around Three Nunavut Communities." PhD diss., University of Toronto, 2007.

———, ed. *Inuit Siku (Sea Ice) Atlas*. sikuatlas.ca.

Langston, Nancy. "Thinking like a Microbe: Borders and Environmental History." *Canadian Historical Review* 95 (2014): 592–603.

Laugrand, Frédéric. *Mourir et renaître: La réception du christianisme par les Inuit de l'Arctique de l'Est canadien (1890–1940)*. Quebec: Les Presses de l'Université Laval, 2002.

Laugrand, Frédéric, and Jarich Oosten. *Hunters, Predators, and Prey: Inuit Perceptions of Animals*. New York: Berghahn Books, 2014.

———. *Inuit Shamanism and Christianity: Transitions and Transformations in the Twentieth Century*. Montreal: McGill-Queen's University Press, 2009.

———. *The Sea Woman: Sedna in Inuit Shamanism and Art in the Eastern Arctic*. Fairbanks: University of Alaska Press, 2008.

———. "*Quviasukvik*: The Celebration of an Inuit Winter Feast in the Central Arctic." *Journal de la société des américanistes* 88 (2002): 203–25.

Laugrand, Frédéric, Jarich Oosten, and David Serkoak. "The Saddest Time of My Life: Relocating the Aharmiut from Ennandai Lake (1950–1958)." *Polar Record* 46:237 (2010): 113–35.

Laugrand, Frédéric, Jarich Oosten, and François Trudel, eds. *Apostle to the Inuit: The Journals and Ethnographic Notes of Edmund James Peck: The Baffin Years, 1894–1905*. Toronto: University of Toronto Press, 2006.

Lears, T. J. Jackson. *No Place of Grace: Antimodernism and American Culture, 1880–1920*. Chicago: University of Chicago Press, 1994.

LeFlouria, Talitha. *Chained in Silence: Black Women and Convict Labor in the New South*. Chapel Hill: University of North Carolina Press, 2015.

Lewis-Jones, Huw W. G. "'Heroism Displayed': Revisiting the Franklin Gallery at the Royal Naval Exhibition, 1891." *Polar Record* 41 (2005): 185–203.

Limerick, Patricia Nelson. *The Legacy of Conquest: The Unbroken Past of the American West*. New York: Norton, 1987.

Litwack, Leon. *Trouble in Mind: Black Southerners in the Age of Jim Crow*. New York: Knopf, 1998.

Loo, Tina. "Political Animals: Barren-Ground Caribou and Managers in a 'Post-Normal' Age." *Environmental History* (March 2017): 1–27.

———. "Disturbing the Peace: Environmental Change and the Scales of Justice on a Northern River." *Environmental History* 12:4 (2007): 895–919.

Loomis, Chauncey C. *Weird and Tragic Shores: The Story of Charles Francis Hall, Explorer*. Lincoln: University of Nebraska Press, 1991.

———. "The Arctic Sublime." In *Nature and the Victorian Imagination*. Ed. U. C. Knoepflmacher and G. B. Tennyson. Berkeley: University of California Press, 1977: 95–112.

Low, A. P. *Report on the Dominion Government Expedition to Hudson Bay and the Arctic Islands on Board the D. G. S.* Neptune, *1903–1904*. Ottawa: Government Printing Bureau, 1906.

Lowenstein, Tom. *Ultimate Americans: Point Hope, Alaska: 1826–1909*. Fairbanks: University of Alaska Press, 2008.

Luciano, Dana. *Arranging Grief: Sacred Time and the Body in Nineteenth-Century America*. New York: New York University Press, 2007.

Lynch, Kevin. *What Time Is this Place?* Cambridge, MA: MIT Press, 1972.

M'Donald, Alexander. A Narrative of Some Passages in the History of Eenoolooapik: A Young Esquimaux, Who Was Brought to Britain in 1839. Edinburgh: Fraser, 1841.

MacDonald, John. *The Arctic Sky: Inuit Astronomy, Star Lore, and Legend*. Toronto: Royal Ontario Museum, 1998.

Macfarlane, Robert. *Landmarks*. Toronto: Penguin Books, 2016.

Maclaren, Ian. "The Aesthetic Map of the North, 1845–1859." *Arctic* 38 (1985): 89–103.

———. "The Limits of the Picturesque in British North America." *Journal of Garden History* vol. 5, no. 1 (1985): 97–111.

———. "Retaining Captaincy of the Soul: Response to Nature in the First Franklin Expedition." *Essays on Canadian Writing* no. 28 (Spring 1984): 57–92.

Maddox, Lucy. *Removals: Nineteenth-Century American Literature and the Politics of Indian Affairs*. New York: Oxford University Press, 1991.

Markham, Sir Clement Roberts, ed. "The Arctic Expedition." *Geographical Magazine* 4 (1877).

Marino, Elizabeth. "Environmental Migration in a Climate of Change: The Anthropology of Social Vulnerability, Disaster, and Justice." In *Environmental Anthropology: Future Directions*. Ed. H. Kopnina and E. Ouimet. New York: Routledge, 2013.

———. "The Long History of Environmental Migration: Assession Vulnerability Construction and Obstacles to Successful Relocation in Shishmaref, Alaska." *Global Environmental Change* 22 (2012): 374–81.

Martin, Charles D. *The White African American Body: A Cultural and Literary Exploration.* New Brunswick, NJ: Rutgers University Press, 2002.

Martin, Keavy. *Stories in a New Skin: Approaches to Inuit Literature.* Winnipeg: University of Manitoba Press, 2012.

Mary-Rousselière, Guy. *Qitdlarssuaq: The Story of a Polar Migration.* Translated by Alan Cooke. Winnipeg: Wuerz Publishing Ltd., 1991.

Matt, Susan J. *Homesickness: An American History.* New York: Oxford, 2011.

———. "You Can't Go Home Again: Homesickness and Nostalgia in U.S. History," *Journal of American History* 94 (2007): 459–87.

McClintock, Francis. *The Voyage of the Fox in the Arctic Seas: A Narrative of the Discovery of the Fate of Sir John Franklin and His Companions.* London: John Murray, 1859.

McCorristine, Shane. "Searching for Franklin: A Contemporary Canadian Ghost Story." *British Journal of Canadian Studies* 26 (2013): 39–58.

McCrossen, Alexis. *Holy Day, Holiday: The American Sunday.* Ithaca, NY: Cornell University Press, 2000.

McGhee, Robert, and Canadian Museum of Civilization. *The Last Imaginary Place: A Human History of the Arctic World.* New York: Oxford University Press, 2005.

McGrath, Janet Tamalik. "*Isumaksaqsiurutigijakka*: Conversations with Aupilaarjuk Towards a Theory of Inuktitut Knowledge Renewal." PhD diss., Carleton University, 2011.

Melville, Herman. *Moby-Dick; or, The Whale.* New York: Norton, 1892 (1851).

Mitchell, Edward D., and Randall R. Reeves, "Factors Affecting Abundance of Bowhead Whales Balaena Mysticetus in the Eastern Arctic of North America, 1915–1980." *Biological Conservation* 22 (1982): 59–78.

Mulder, J., and C. H. Stuart-Harris. "Influenzal Pneumonia: Causation and Treatment." *Bulletin of the World Health Organization* 8 (1953): 743–53.

Müller-Wille, Ludger, ed. *Franz Boas among the Inuit of Baffin Island, 1883–1884.* Trans. William Barr. Toronto: University of Toronto Press, 1998.

Müller-Wille, Ludger, and Bernd Gieseking, eds. *Inuit and Whalers on Baffin Island through German Eyes: Wilhelm Weike's Arctic Journal and Letters (1883–84).* Trans. William Barr. Montreal: Baraka Books, 2011.

Nakasuk, Saullu, et al. *Interviewing Inuit Elders: Introduction.* Ed. Jarich Oosten and Frédéric Laugrand. Iqaluit: Nunavut Arctic College, 1999.

Nash, Linda. *Inescapable Ecologies: A History of Environment, Disease, and Knowledge.* Berkeley: University of California Press, 2006.

———. "The Changing Experience of Nature: Historical Encounters with a Northwest River." *Journal of American History* 86 (2000): 1600–1629.

National Maritime Digital Library. *American Offshore Whaling Voyages: A Database.* Online: http://nmdl.org/aowv/whindex.cfm.

Nickerson, Sheila B. *Midnight to the North: The Untold Story of the Woman Who Saved the Polaris Expedition.* New York: Putnam, 2002.

Nagy, Murielle, ed. "Intellectual Property and Ethics." *Études/Inuit/Studies* 35 (2011): 7–342.

New Bedford Whaling Museum. "Whaling Crew List Database." Online: https://www.whalingmuseum.org/online_exhibits/crewlist/

Norling, Lisa. *Captain Ahab Had a Wife: New England Women and the Whalefishery, 1720–1870.* Chapel Hill: University of North Carolina Press, 2000.

Nourse, J. E., ed. *Narrative of the Second Arctic Expedition Made by Charles F. Hall.* Washington, DC: Government Printing Office, 1879.

Nowicki, Mel. "Rethinking Domicide: Towards an Expanded Critical Geography of Home." *Geography Compass* 8:11 (2014): 785–95.

Nunavut Wildlife Management Board. *Final Report of the Inuit Bowhead Knowledge Study*. Iqaluit: Nunavut Wildlife Management Board, 2000.

Nuttall, Mark. "Anticipation, Climate Change, and Movement in Greenland." *Études/Inuit/Studies* 34:1 (2010): 21–37.

———. "The Name Never Dies: Greenlandic Inuit Ideas of the Person." In *Amerindian Rebirth: Reincarnation Belief among North American Indians and Inuit*. Ed. A. Mills and R. Slobodin. Toronto: University of Toronto Press, 1994.

———. *Arctic Homeland: Kinship, Community, and Development in Northwest Greenland*. London: Belhaven, 1992.

Nuttall, Mark, ed. *Encyclopedia of the Arctic*. New York: Routledge, 2005.

Obed, Natan. "The True North—or rather True Inuit Nunangat—Strong and Free." Canadian Broadcasting Corporation, June 23, 2017. http://www.cbc.ca/2017/the-true-north-or-rather-true-inuit-nunangat-strong-and-free-1.4173047.

Oosten, Jarich, and Frédéric Laugrand. "*Qaujimajatuqangit* and Social Problems in Modern Inuit Society: An Elders Workshop on *Angakkuuniq*." *Études/Inuit/Studies* 26:1 (2002): 17–44.

Ootoova, Ilisapi, et al. *Perspectives on Traditional Health*. Ed. Michèle Therrien and Frédéric Laugrand. Iqaluit: Nunavut Arctic College, 2001.

Outerbridge, Joseph. "My Voyage to Cumberland Inlet." *Atlantic Advocate* 59 (1969): 50–68.

Pagh, Nancy. *At Home Afloat: Women on the Waters of the Pacific Northwest*. Calgary: University of Calgary Press, 2001.

Parr, Joy. *Sensing Changes: Technologies, Environments, and the Everyday, 1953–2003*. Vancouver: UBC Press, 2010.

Parry, Richard. *Trial by Ice: The True Story of Murder and Survival on the 1871 Polaris Expedition*. New York: Ballantine, 2001.

Partridge, Shannon, ed. *Niurrutiqarniq: Trading with the Hudson's Bay Company*. Iqaluit: Nunavut Arctic College, 2009.

Peck, Robert. "The Art of the Arctic: British Painting in the Far North." *Journal for Maritime Research* 14 (2012): 67–93.

Petersen, Robert. "Om qivittut—fjeldgængere." *Grønland* 5 (2006): 203–15.

———. "Eskimoernes Sidste Indvandring fra Canada till Grønland." *Tidsskriftet Grønland* 12 (1964): 373–87.

———. "The Last Eskimo Immigration into Greenland." *Folk: Dansk Etnografisk Tidsskrift* 4 (1962): 95–110.

Petrone, Penny, ed. *Northern Voices: Inuit Writing in English*. Toronto: University of Toronto Press, 1992.

Pielou, E. C. *A Naturalist's Guide to the Arctic*. Chicago: University of Chicago Press, 1994.

Piper, Liza. "Chesterfield Inlet, 1949, and the Ecology of Epidemic Polio." *Environmental History* 20 (2015): 671–98.

———. *The Industrial Transformation of Subarctic Canada*. Vancouver: UBC Press, 2009.

Piper, Liza, and John Sandlos, "A Broken Frontier: Ecological Imperialism in the Canadian North," *Environmental History* 12.4 (2007): 759–95.

Pitseolak, Markosie. *Markosie Pitseolak's Real Life Stories*. Trans. Andrew Dialla. Pangnirtung, 1973. Original in possession of Rosee and Pauloosie Veevee's family.

Porteous, J. Douglas, and Sandra E. Smith. *Domicide: The Global Destruction of Home*. Montreal: McGill-Queen's University Press, 2001.

Potter, Russell A. *Arctic Spectacles: The Frozen North in Visual Culture, 1818–1870*. Seattle: University of Washington Press, 2007.

Price, Jackie. "The Arctic Is My Home: Affirming the Art of Inuit Governance." *Fuse* 35:2 (2012): 2–3.

Public Broadcasting Corporation. "Interview with Michael Robinson." *American Experience*, Season 23, Episode 5. http://www.pbs.org/video/american-experience-interview-with-michael-robinson/.

Qikiqtani Inuit Association. *Illiniarniq: Schooling in Qikiqtaaluk*. Qikiqtani Truth Commission Thematic Reports and Special Studies, 1950–1975. Iqaluit: Inhabit Media, 2013.

———. *Nuutauniq: Moves in Inuit Life*. Qikiqtani Truth Commission Thematic Reports and Special Studies, 1950–1975. Iqaluit: Inhabit Media, 2013.

———. *The Official Mind of Canadian Colonialism*. Iqaluit: Inhabit Media, 2013.

———. *Paliisikkut: Policing in Qikiqtaaluk*. Qikiqtani Truth Commission Thematic Reports and Special Studies, 1950–1975. Iqaluit: Inhabit Media, 2013.

———. *Qikiqtani Truth Commission: Community Histories 1950–1975: Pangnirtung*. Iqaluit: Inhabit Media, 2013.

———. *Qimmiliriniq: Inuit Sled Dogs in Qikiqtaaluk*. Qikiqtani Truth Commission Thematic Reports and Special Studies, 1950–1975. Iqaluit: Inhabit Media, 2013.

———. *QTC Final Report: Achieving Saimaqatigiingniq*. Iqaluit: Inhabit Media, 2013.

Qitsualik, Rachel Attituq. "Is It 'Eskimo' or 'Inuit'?" *Indian Country Today Media Network*. November 11, 2004. http://indiancountrytodaymedianetwork.com/2004/02/11/qitsualik-it-eskimo-or-inuit-89951.

———. "Cannibal." *Nunatsiaq News*, August 1999 (four-part series).

———. "Ilira." Pts. 1, 2, 3. *Nunatsiaq News*, November 12, 19, 26, 1998.

———. "Nunataaq: Nunavut as the Inuit Promised Land." *Nunatsiaq News*, October 16, 1998.

Rasmussen, Derek, and Tommy Akulukjuk. "'My Father Was Told to Talk to the Environment First Before Anything Else': Arctic Environmental Education in the Language of the Land." In *Fields of Green: Restorying Culture, Environment, and Education*. Ed. Marcia McKenzie, Paul Hart, Heesoon Bai, and Bob Jickling. New York: Hampton Press, 2009.

Rasmussen, Knud. *The People of the Polar North: A Record*. Trans. and ed. G. Herring. London: Kegan Paul, Trench, Trübner, 1908.

Reilly, S. B., et al. *Balaena mysticetus*. The IUCN Red List of Threatened Species 2012: e.T2467A17879018. http://dx.doi.org/10.2305/IUCN.UK.2012.RLTS.T2467A17879018.en.

Rice, George. "Journal: Sergeant Rice." Published online by American Experience, Public Broadcasting Service. Currently offline but accessible through Wayback Machine: https://web.archive.org/web/20110205015023/http://www.pbs.org/wgbh/americanexperience/features/primary-resources/greely-rice-journal/.

Richter, Daniel K. *Facing East from Indian Country: A Native History of Early America*. Cambridge, MA: Harvard University Press, 2001.

Riffenburgh, Beau. *The Myth of the Explorer: The Press, Sensationalism, and Geographical Discovery*. New York: Oxford University Press, 1994.

Ritvo, Harriet. "Going Forth and Multiplying: Animal Acclimatization and Invasion. President's Lecture." *Environmental History* 17:2 (2012): 404–14.

Robinson, Michael F. *The Coldest Crucible: Arctic Exploration and American Culture*. Chicago: University of Chicago Press, 2006.

Rosenkrantz, Barbara Gutmann, ed. *From Consumption to Tuberculosis: A Documentary History*. New York: Garland, 1994.

Rosenow, Michael K. *Death and Dying in the Working Class, 1865–1920*. Champaign: University of Illinois Press, 2015.

Ross, W. Gillies. "The Type and Number of Expeditions in the Franklin Search, 1847–1859." *Arctic* 55 (2002): 57–69.

———. *This Distant and Unsurveyed Country: A Woman's Winter on Baffin Island, 1857–1858*. Montreal: McGill-Queen's University Press, 1997.

———. *Arctic Whalers, Icy Seas: Narratives of the Davis Strait Whale Fishery*. Toronto: Irwin, 1985.

———. "The Annual Catch of Greenland (Bowhead) Whales in Waters North of Canada 1719–1915: A Preliminary Compilation." *Arctic* 32 (1979): 91–121.

———. "Canadian Sovereignty in the Arctic: The "Neptune" Expedition of 1903–04." *Arctic* 29 (1976): 87–104.

Rowley, Susan. "Eenoolooapik," *Dictionary of Canadian Biography Online*. Toronto: University of Toronto Press, 2005. http://www.biographi.ca.

Rugh, David J., and Kim E. W. Shelden. "Bowhead Whale." In *Encyclopedia of Marine Mammals*. 2nd ed. Ed. William F. Perrin, Bernd Würsig, and J. G. M. Thewissen. Burlington, MA: Academic Press, 2009: 131–33.

Russell, Sharman Apt. *Hunger: An Unnatural History*. New York: Basic Books, 2006.

Ryan, Mary P. *Cradle of the Middle Class: The Family in Oneida County, New York, 1790–1865*. Cambridge: Cambridge University Press, 1983.

Ryan, James R. "'Our Home on the Ocean': Lady Brassey and the Voyages of the *Sunbeam*, 1874–1887." *Journal of Historical Geography* 32 (2006): 579–604.

Ryan, Jerry. *The Forgotten Aquariums of Boston*. 3rd rev. ed. Pascoag, RI: Finley Aquatic Books, 2011 (2002).

Said, Edward. *Orientalism*. New York: Vintage Books, 1978.

Saladin d'Anglure, Bernard. "La toponymie réligieuse et l'appropriation symbolique du territoire par les Inuit du Nunavik et du Nunavut." *Études/Inuit/Studies* 28 (2004): 107–31.

———. "Brother Moon, Sister Sun, and the Direction of the World: From Arctic Cosmography to Inuit Cosmology." In *Circumpolar Religion and Ecology: An Anthropology of the North*, 187–212. Ed. Takashi Irimoto and Takako Yamada. Tokyo: University of Tokyo Press, 1994.

Sandiford, Mark. *Qallunaat! Why White People Are Funny*. Film. Beachwalker Films/National Film Board, 2006.

Sandlos, John. *Hunters at the Margin: Native People and Wildlife Conservation in the Northwest Territories*. Vancouver: UBC Press, 2007.

Savours, Ann. *The North West Passage in the Nineteenth Century: Perils and Pastimes of a Winter in the Ice*. London: Hakluyt Society, 2003.

———. *The Search for the North West Passage*. New York: Palgrave, 1999.

Saxon, A. H. *P. T. Barnum: The Legend and the Man*. New York: Columbia University Press, 1989.

Schivelbusch, Wolfgang. *The Railway Journey: The Industrialization of Time and Space in the Nineteenth Century*. Berkeley: University of California Press, 1986.

Schledermann, Peter. *Voices in Stone: A Personal Journey into the Arctic Past*. Calgary: Arctic Institute of North America, 1996.

Scoresby, William. *An Account of the Arctic Regions with a History and Description of the Northern Whale-Fishery*. 2 vols. Edinburgh: Archibald Constable, 1820.

Scott, James. *The Art of Not Being Governed: An Anarchist History of Upland Southeast Asia*. New Haven, CT: Yale University Press, 2009.

Simonsen, Jane E. *Making Home Work: Domesticity and Native American Assimilation in the American West, 1860–1919*. Chapel Hill: University of North Carolina Press, 2006.

Simpson, Audra. *Mohawk Interruptus: Political Life Across the Borders of Settler States*. Durham, NC: Duke University Press, 2014.

Smith, Mick, Joyce Davidson, and Laura Cameron, eds. *Emotion, Place and Culture*. Burlington, VT: Ashgate, 2009.

Smith, Tim D., Randall R. Reeves, Elizabeth A. Josephson, and Judith N. Lund. "Spatial and Seasonal Distribution of American Whaling and Whales in the Age of Sail." *PLoS ONE* 7 (4): e34905. doi:10.1371/journal.pone.0034905.

Sonne, Birgitte, ed. *Grønlandske fortællinger*. 2004. http://arktiskinstitut.dk/da/samlinger/groenlandske-fortaellinger/.

———. "The Acculturative Role of Sea Woman: Early Contact Relations between Inuit and Whites as Revealed in the Origin Myth of the Sea Woman." *Meddelelser om Grønland, Man and Society* 13. Copenhagen: Danish Polar Centre, 1990.

Spalding, Alex. *A Multi-Dialectical Outline Dictionary with an Aivilingmiutaq Base*. Iqaluit: Nunavut Arctic College, 1998.

Spink, Wesley W. *Infectious Diseases: Prevention and Treatment in the Nineteenth and Twentieth Centuries*. Minneapolis: University of Minnesota Press, 1978.

Spufford, Francis. *I May Be Some Time: Ice and the English Imagination*. London: Faber and Faber, 1996.

Stairs, Arlene. "Self Image, World-Image: Speculations on Identity from Experiences with Inuit." *Ethos* 20:1 (1992): 116–26.

Starbuck, Alexander. *History of the American Whale Fishery, from Its Earliest Inception to the Year 1876*. Washington: Government Printing Office, 1876.

Stenton, Douglas R. "The Adaptive Significance of Caribou Winter Clothing for Arctic Hunter-Gatherers." *Etudes/Inuit/Studies* 15:1 (1991): 3–28.

———. "Caribou Population Dynamics and Thule Culture Adaptations on Southern Baffin Island, NWT." *Arctic Anthropology* 28:2 (1991): 15–43.

Stern, Pamela. "Upside Down and Backwards: Time Discipline in a Canadian Inuit town." *Anthropologica* 45:1 (2003): 147–61.

Stern, Pamela R., and Lisa Stevenson, eds. *Critical Inuit Studies: An Anthology of Contemporary Arctic Ethnography*. Lincoln: University of Nebraska Press, 2006.

Stevenson, Marc. *Inuit, Whalers, and Cultural Persistence: Structure in Cumberland Sound and Central Inuit Social Organization*. Toronto: Oxford University Press, 1997.

———. "Kekerten: Preliminary Archaeology of an Arctic Whaling Station." Yellowknife, NT: Prince of Wales Heritage Centre, 1984.

Stuckenberger, Nicole. *Thin Ice: Inuit Traditions within a Changing Environment*. Hanover, NH: Dartmouth College, 2007.

Stuhl, Andrew. "Cold Places: Movement, Knowledge, and Time." *Journal of Environmental Studies and Sciences* 6:4 (2016): 779–82.

———. *Unfreezing the Arctic: Science, Colonialism, and the Transformation of Inuit Lands*. Chicago: University of Chicago Press, 2016.

Tester, Frank, Paule McNicoll, and Quyen Tran. "Structural Violence and the 1962–63 Tuberculosis Epidemic in Eskimo Point, N.W.T." *Études/Inuit/Studies* 36 (2012): 165–85.

Tester, Frank, Paule McNicoll, and Peter Irniq. "Writing for Our Lives: The Language of Homesickness, Self-Esteem and the Inuit TB 'epidemic.'" *Études/Inuit/Studies* 25 (2001): 121–40.

Tester, Frank, and Peter Kulchyski. *Tammarniit (Mistakes): Inuit Relocation in the Eastern Arctic, 1939–63*. Vancouver: UBC Press, 1994.

Thompson, E. P. "Time, Work-Discipline, and Industrial Capitalism." *Past and Present* 38 (December 1967): 56–97.

Thorpe, Natasha, Kitikmeok Elders, et al. *Thunder on the Tundra: Inuit Qaujimajatuqangit of the Bathurst Caribou*. Victoria, BC: Tuktu and Nogak Project, 2002.

Thrush, Coll. *Indigenous London: Native Travellers at the Heart of Empire*. New Haven, CT: Yale University Press, 2016.

———. "The Iceberg and the Cathedral: Encounter, Entanglement, and Isuma in Inuit London." *Journal of British Studies* 53 (2014): 59–79.

Todd, A. L. *Abandoned: The Story of the Greely Arctic Expedition, 1881–1884*. Fairbanks: University of Alaska Press, 2001.

Trott, Christopher G. "The Gender of the Bear." *Etudes/Inuit/Studies* 30:1 (2006): 89–109.

———. "'Reading' Conversion Narratives: Textualized Oralities." Paper presented at the Inuit Studies conference, Paris, October 2006.

———. "'There Have Been Grave Disappointments': Reading Inuit Agency through Missionary Texts." Presentation for the Native Studies Department Colloquium Series, University of Manitoba, January 18, 2006.

———. "Ilagiit and Tuqłuraqtuq: Inuit Understandings of Kinship and Social Relatedness." Conference paper presented at *First Nations, First Thoughts*, Centre of Canadian Studies, University of Edinburgh, 2005.

———. "The Dialectics of 'Us' and 'Other': Anglican Missionary Photographs of the Inuit." *American Review of Canadian Studies* 31:1 (2001): 171–90.

———. "The Cannibalism Theme among the Inuit: A Case Study in Inversions." *Igitur* VI–VII:1187–2122 (1995): 17–37.

Tuan, Yi-Fu. "Home." In *Patterned Ground: Entanglements of Nature and Culture*, 164-65. Ed. Stephan Harrison, Steve Pile, and N. J. Thrift. London: Reaktion Books, 2004.

———. *Space and Place: The Perspective of Experience*. Minneapolis: University of Minnesota Press, 1977.

Twain, Mark, et al. *Mark Twain's Letters: 1872–1873*. Berkeley: University of California Press, 1997.

Tyson, George E. *The Cruise of the Florence: or, Extracts from the Journal of the Preliminary Arctic Expedition of 1877-78*. Washington, DC: J. J. Chapman, 1879.

United States Navy Department. Annual Report of the Secretary of the Navy on the Operations of the Department for the Year 1873. Washington, DC: Government Printing Office, 1873.

United States Army Center of Military History. "Medal of Honor Recipients, Interim Awards, 1920–1940." http://www.history.army.mil/html/moh/interim1920-40.html.

Urness, James. *Twenty-Five Brave Men: Tales from an Arctic Journey*. Ed. Donald Kvamme. Tucson: Wheatmark, 2015.

Valenčius, Conevery Bolton. *The Health of the Country: How American Settlers Understood Themselves and Their Land*. New York: Basic Books, 2004.

Vaughan, Alden T. *Transatlantic Encounters: American Indians in Britain, 1500–1776*. New York: Cambridge University Press, 2006.

Von Finckenstein, Maria, ed. *Nuvisavik: The Place Where We Weave*. Montreal: McGill-Queen's University Press, 2002.

Wachowich, Nancy, Apphia Agalakti Awa, Rhoda Kaukjak Katsak, and Sandra Pikujak Katsak. *Saqiyuq: Stories from the Lives of Three Inuit Women*. Montreal: McGill-Queen's University Press, 1999.

Wakeham, William. *Report of the Expedition to Hudson Bay and Cumberland Gulf in the Steamship* Diana *under the Command of William Wakeham, Marine and Fisheries Canada, in the Year 1897.* Ottawa: Queen's Printer, 1898.

Warf, Barney. *Time-Space Compression: Historical Geographies.* London: Routledge, 2008.

Wenzel, George W. *Animal Rights, Human Rights: Ecology, Economy, and Ideology in the Canadian Arctic.* Toronto: University of Toronto Press, 1991.

Weslawski, Jan Marcin, and Joanna Legezynska. "Chances for Arctic Survival: Greely's Expedition Revisited." *Arctic* 55:4 (2002): 373–79.

White, Richard. *The Organic Machine.* New York: Hill and Wang, 1995.

Williams, Kidada E. *They Left Great Marks on Me: African American Testimonials of Racial Violence from Emancipation to World War I.* New York: New York University Press, 2012.

Wilson, Eric G. *The Spiritual History of Ice: Romanticism, Science, and the Imagination.* New York: Palgrave Macmillan, 2003.

Wilson, Lisa. *Ye Heart of a Man: The Domestic Life of Men in Colonial New England.* New Haven, CT: Yale University Press, 2000.

Womack, Craig. *Red on Red: Native American Literary Separatism.* Minneapolis: University of Minnesota Press, 1999.

Worster, Donald. *Dust Bowl: The Southern Plains in the 1930s.* New York: Oxford University Press, 2004 (1979).

Wright, Carroll D. *Comparative Wages, Prices, and Cost of Living.* Boston: Wright and Potter, 1889.

Index

NOTE: Page numbers followed by *f* indicate a figure.

Aasivak, 114, 115*f*, 116, 131, 140; midwifery services of, 123–24; religious practices of, 121–23, 125, 189n30
acclimatization societies, 164n44
Akpalialuk, 152–53, 195n9
Aksayuk, Etooangat, 61, 111*f*, 112*f*, 140, 188n1, 190n40, 190n53; Christian childhood of, 123–25; daughter of, 119; on the end of commercial whaling, 136, 192n85; on food shortages and hunger, 141–44; home of, 113; official birthdate of, 123, 189n33; stories of the old days of, 110–12, 146–47; on winter, 117
Akulukjuk, Malaya, 116, 119*f*
Albert, 135–36
Americans. *See* Qallunaat (non-Indigenous outsiders)
Ancestors in the Attic, 140–41
Andrews, 11–12, 18, 32, 167n33
angirraq, xx–xxi, 163n30. *See also* home
Angmarlik, 122–23, 129, 189n31, 190n51
Angmarlik, Pauloosie, 46, 134, 166n8, 190n51
Angry Inuk (Arnaquq-Baril), 150
Angutiqjuaq, 61, 110–11
Angutisiak, Jens Edvard, 78–80, 88–94, 100–103, 184nn72–73, 185n121; death of, 102–3, 107, 186nn139–40; flight from Fort Conger by, 91–94; home community of, 88–91, 183n49
Aniiniliak, Evie, 130, 189n23
Ansel Gibbs, 7, 24, 170n111
Antelope, 1–2, 7, 11, 165n1
Arctic: early exploration of, xvi; environmental history of, xxiv–xxvi, 164nn44–46; as home, xiv, xix–xxiii, 32–34; impact of sunlight in, 99–103, 185n112; impact of winter darkness in, 80–87, 90, 99, 182n31; Inuit descriptions of, xiv, 19–20, 56, 111–12, 118–19, 158; Qallunaat stereotypes of, xiv–xv, xx, xxvi, 1–2, 8, 10, 17–20, 32, 78–80, 108–9, 151, 168n63, 169n72, 182n31; saleable resources of, xiv, 5, 28–31, 166n16; seasons in, 2–3, 166n8. *See also* Cumberland Sound; expeditions to the Arctic; Inuit; Lady Franklin Bay expedition; whaling industry
Arnaqoq, 138
Arnaquq, Billy, 145–46
Arnaquq-Baril, Alethea, 150
Atagutsiak, Tipuula Qaapik, 180n165
aurora borealis, 85
authenticity, 44–45, 174n33

Bachelard, Gaston, 87
baleen (whalebone), 5, 28–31, 115, 127, 166n16
Bandwell, William, 4
Barnum, Phineas Taylor, 40–41, 174n36
Barnum's American Museum, 40–45, 173n10, 179n136
Barron, William, 166n16
Bates, Ambrose, 15–19, 24, 30–31, 153, 173n6
Baudrillard, Jean, 87
Bek, Adam, 16
beluga whales, 31, 40, 128–31, 191n59
Bessuell, Emil, 23
Bilby, Julian, 129*f*, 143, 189n31
Black Eagle, 24, 169n102
Blacklead Island. *See* Uummannarjuaq
blubber, 5, 28–31, 166n16
Boas, Franz, xix, 14, 190n40; on Inuit homes, 168n46; on Inuit naming traditions, 20

bowhead whales (*Balaena mysticetus*), xxii, xxvii, 27–31, 114–15; decline of, xix–xx, 31, 127–29, 136–37, 144–45; legal hunting in Nunavut of, 145–48, 194nn117–18. *See also* whaling industry
bowhead whale tongue, 141–42
Bradbury, Bettina, 71
Brainard, David, 83, 87, 93, 97, 99–100, 102–7
Brevoort, Henry L., 69–70
Briggs, Jean, 178n123
Brody, Hugh, xxi
Brown, John, 26
Budington, James Monroe, 6, 10–11
Budington, Sarah, ix, 50, 54–57, 59, 64–65, 71–74, 153
Budington, Sidney, 34*f*, 153; on alleviation of scurvy, 26; on the Arctic and Inuit, xiv, 33, 161n7; Hall and, 50, 52–54, 59–60, 65–66; on overwintering, 166n16; voyages to Cumberland Sound of, 6–7, 9, 11*f*
Busch, Briton Cooper, 167n19

Cameron, Emilie, xxiv, 164n40
Canadian government policies: on assignment of surnames, 140–41; on creation of permanent settlements, xix, xx, 52, 58, 134, 149–51, 153, 195n10; Inuit resistance to, 113, 150–51, 194n3; on RCMP in Cumberland Sound, 139–40, 192n91
cannibalism, 9–10, 106–7
Cardno, David, 115–17, 167n37
caribou hunting, 131–35, 191nn71–72
CGS Arctic, 189n33
Chicago World's Columbian Exposition of 1893, 108
Children Dancing to the Whaler's Jig (Ishulutaq), 21*f*
Christiansen, Thorleif Frederik, 89–93, 101–2
Clark, G. V., 192n85
Clyde River, 150, 194n3
Coleman, Patrick, 61–62
Collignon, Béatrice, xxi
colonialism, xiv–xvi, xxii–xxv, 61, 74, 149–51, 154–55; Inuit labor and, 130–31; time and, xxvi. *See also* expeditions to the Arctic; whaling industry
Comer, George, 33
Connell, Maurice, 103, 105, 108
Copp, John Joseph, 67, 179n131
Cross, Sergeant, 93, 96, 98
Cruikshank, Julie, xxiv
Cumberland Sound, xvi–xx, 108, 181n4; beluga herding in, 128–31, 191n59; early exploration of, xvi; Hudson's Bay Company in, 136–38, 153; maps of, xvii*f*, xviii*f*; Pangnirtung as permanent settlement for, xix, 149–51, 158–60; place names in, xvi, xviii*f*, xxi, 20, 110, 112, 126, 157, 160, 163n19, 166n11; RCMP post in, 139–40, 192n91; scientific research in, xix; whaling industry of, xvi–xx, 31–34, 110–11, 114–15, 127–29, 163n22. *See also* Inuit in Cumberland Sound; Qallunaat in Cumberland Sound; seasons in Cumberland Sound
Cutting, James Ambrose, 174n36, 179n136

Daniel Webster, 7–8, 23–24, 169n102
darkness, 80–87, 90, 99, 182n31
David, Robert G., 182n31
Davidson (cooper), 113–14
Davis, John, xvi
Davis Strait whaling industry, xvii, 163n22. *See also* Cumberland Sound
daylight, 99–103, 185n112
Dialla, Andrew, xiv, 107, 139*f*, 140–42, 158, 188n1, 190n40
Dialla, Daisy, 122, 128, 137–40, 144
Dialla, David, 138–39
Dialla, Joanasie, 132, 138–40
Dick, Lyle, xxiv, 79
Dolin, Eric Jay, 166n6
Dunning, Norma, 162n13
Dutton, Warren, 8–9
Duval, William Sivutiksaq, 34, 152, 154
dwellings. *See* housing

early fall (*ukiaksaaq*), 2, 7–10, 135–41
early spring (*upingaksaaq*), 3, 22–27, 120–25
Easonian, 135–36, 192n85
Ebierbing, Hannah and Ipiirvik, 35–76, 153, 173n6; adaptation to American life of, 47–48, 67–69, 72, 75–76; adopted daughter of, 65, 66*f*, 70; American "family" of, 64–65, 178n123; Arctic expeditions with Hall of, 35–36, 62–69, 173n2; conceptions of home of, 55–59; Connecticut home of, 65–67, 179n131; deaths of children of, 50–57, 65, 70, 73; drawings by, 56*f*, 57*f*; as exhibits of difference, 40–50; finances of, 65–69, 179n131, 179n136; Hall's authoritarianism towards, 59–65, 178n116; illnesses of, 50–54, 70–74, 176n77, 180n165; Inuktitut names of, 38*f*, 173n6; Ipiirvik's final return to the Arctic, 75, 181n178; itinerant lives of, 50–59; language skills of, 44, 54; photographs of, 37*f*, 55*f*, 63*f*; traditional clothing of, 37*f*, 42, 43*f*, 47–48; travel abroad of, 37–38, 48; written accounts of, 38
Ebierbing, Sylvia Grinnell (Isigaittuq), 65, 66*f*, 70
Eenoolooapik (Inuluapik), xvi–xvii, 47, 48, 153
Elison, Joseph, 98–100, 103
Ellesmere Island, 79–80, 88, 91, 181n4. *See also* Lady Franklin Bay expedition
environmental history, xxiv–xxvi, 164n43; consideration of home in, xxiv–xxv, 149–55; understandings of time in, xxv–xxvi, 39

INDEX

Ernest William, 138
Eskimo/Esquimaux (as term), xi–xii. *See also* Inuit
Evans, Thomas, 166n12
Evic, Qatsu, 113–19, 121, 123, 188n5, 189n33; on caribou hunting, 131–32; childhood *qammaq* of, 118–19; on home, 131, 191n65; marriage of, 118, 189n18; on whaling, 127; on winter, 117–18
expeditions to the Arctic, 1–2; as colonialist incursions, xv–xvi, xxiv, 151; of Franklin, xiv, xxvii, 1–2, 7, 46; of Hall, 7, 38, 55, 58, 65–69, 173n2; to Lady Franklin Bay, xix, 78–109; of *Polaris*, xi–xv, xviif, xxvi–xxviii, 33, 69, 74–75, 161n2, 161n5, 183n54; sensationalizing of, xiv–xv, 7, 104, 106–9, 161n7, 162n11. *See also* whaling industry

fall. *See* early fall (*ukiaksaaq*); late fall (*ukiaq*)
Faulkner, Joseph, 168n50
Feeding My Family, 150
First International Polar Year, xix, 78, 108, 181n4
Fisher, Samuel, 9–10
Fisheries Act (Canada), 145
floe-edge whaling, 27–31, 125–31, 190n40, 190nn42–43
Flora, Janne, 91
fox trapping, 136–37
Franklin, 11f
Franklin, John, xiv, xxvii, 1–2, 7, 46
Frasier, Charles, 23, 26
Freeman, Minnie Aodla, 162n13
frostbite, 11, 14, 26, 94, 98, 103, 120

Gardiner, Hampden, 78–80, 82f, 103, 105, 109; on Angutisiak's work, 184n72; on the expedition's retreat, 96; health challenges of, 81–82; scientific training of, 81; on winter in the High Arctic, 81–87, 92, 182n31
George Henry, 35, 36f, 50
governance. *See* Canadian government policies
Grant, Ulysses S., 65
Greely, Adolphus, 77–89, 92–108, 159; on the Arctic environment, 86, 185n118; on Inuit expedition members, 92–94, 184nn72–73; leadership style of, 96, 98, 101, 186n128; military service of, 181n1. *See also* Lady Franklin Bay expedition
Greenshield, Edgar, 113–14
Grinnell, Henry, 58, 59–60, 64–69

Hall, Charles Francis, xix, 159; Arctic expeditions of, 7, 38, 55, 58, 65–69, 173n2; authoritarian temperament of, 59–65, 68, 178n116; descriptions of home of, 58; family of, 64, 178n118; Grinnell's sponsorship of, 58–60, 64–69; on Inuit disease, 74, 180n170; on Inuit society, 58–59; lectures and tours of, 42, 44–50, 52–54, 59, 64, 65, 67

Hannah Ebierbing. *See* Ebierbing, Hannah and Ipiirvik
Hantzsch, Bernhard, 16, 132, 190n40
Hart, Ira F., 52
Hayes, Isaac Israel, 83
Heimdal, 189n33
Hendrik, Hans, 49, 74–75, 89–91, 161n5, 169n72, 183n54
High Arctic. *See* Lady Franklin Bay expedition
home, xiv–xv, 78–80, 149–55; acceptance of uncertainty and, 147–48; contrasting visions of, xix–xxiii, 149–50, 154–55, 163n30; environmental history and, xxiv–xxv, 164nn44–46; in government-created permanent settlements, xix, xx, 52, 58, 134, 149–51, 153, 195n10; Inuit conceptions of, xix–xxiii, 35–36, 41, 55–59, 86, 113, 163n30, 164n46, 177n87; "memoryscapes" of, 20, 169n80; as political construct, xv–xvi; Qallunaat conceptions of, 15–16, 32–34, 52, 77, 149–50; Qallunaat stereotypes of the Arctic and, xiv–xv, xx, xxvi, 1–2, 8, 10, 17–20, 32, 78–80, 108–9, 151, 168n63, 169n72, 182n31; on traditional lands, xii, 150; understandings of time and, xxv–xxvi, 39; Victorian women's roles in, 15; whaling ships as, 32–34, 172n144
housing: of Inuit living in Pangnirtung, 150; of Inuit living in Qallunaat lands, 56–59, 63–65, 179n131; of the Lady Franklin Bay expedition, 80, 81f, 84f, 85, 93, 95; of Qallunaat in Cumberland Sound, 4, 5f, 12–13, 114, 137f, 188n8; Qallunaat style of, xxi–xxv, 154, 164n38; in traditional Inuit dwellings, xvi, 4, 58, 117–20, 128
Hudson's Bay Company (HBC), xix, 128, 136–38, 153
hunger: experiences of, xxvii, 9–10, 98–109, 131–32, 141–45, 150; Inuit perceptions of, 64, 137, 143–44; Qallunaat perceptions of, 142–43

ice floes: Inuit survival practices on, xi–xiii, 120–21, 161n4; *Polaris* expedition and, xi–xiv, 33, 74, 161n2, 161n5; whaling from, 27–31, 125–31, 190n40, 190nn42–43
igluvigait (igloos), 58, 117, 120
Ilisaqsivik Society, 150, 194n3
Immukke, 189n23
Indigenous rights activism, 150–51
influenza, 70
International Polar Year, xix, 78, 108, 181n4
Inuit: adoption practices among, 65; avoidance of anger and conflict among, 26, 48, 61, 89, 117, 175n50; on cannibalism, 107; childhood games of, 117, 119; Christianity among, xxvi, 16, 39, 90–91, 115, 120–25, 175n42, 189n23, 189n25; clothing made by, 14, 37f, 42, 43f, 47–48, 125, 131–34, 168n50; coastal settlements

Inuit (*continued*)
of, 10, 12, 90, 131–32; conceptions of home among, xix–xxiii, 35–36, 41, 55–59, 163n30, 164n46, 177n87; conceptions of time among, xxv–xxvi, 39, 110–13; core societal values (Inuit Qaujimajatuqangit Principles) of, 151, 195n4; descriptions of homelands by, xvi; diet of, xvi, xxii, xxvii, 4, 22–26, 128, 132, 141–45, 169n72; in Europe and the U.S., xiii–xiv, 37–38, 48, 173n4; family and kinship among, 46, 64, 140–41, 175n42, 178n123; government-assigned surnames and birthdates of, 123, 140–41, 189n33; ice floe survival skills of, xi–xiii, 120–21, 161n4; Indigenous rights activism of, 150–51, 194n3; long-distance travel customs of, xxvi–xxviii, 51–52, 97–98, 131–34, 152, 165nn53–54, 176n65; naming traditions among, xx–xxi, 20, 54, 103, 173n1, 176n75; new diseases among, xxii, 26–27, 71, 124, 132, 143; oral history research among, 157–60; partnering with whalers of, 15–16, 34, 115, 138–39; process of authenticity among, 44–45, 174n33; Qallunaat stereotypes of, xxvi, 14–16, 42–50, 92–94, 142–43, 168n46, 174n18; relocation into permanent settlements of, xix, xx, 52, 58, 134, 149–51, 153, 195n10; seasonal distinctions of, 2–3, 90, 145–46, 166n8; storytelling practices of, 111–12; supernatural beings and shamanic practices of, 20, 90–92, 114, 121–25, 146–47, 183nn60–61; trade and interactions with whalers of, xxi–xxii, 2, 14–15, 21–22, 32–34, 111–17, 137–38; traditional dwellings of, xvi, 4, 58, 117–20, 128; traditional homelands of, xii, 150; travel and living abroad of, 35–38, 41, 48, 55–59, 152–54, 177n87; treatment of illness among, 73, 180n165; uses of whale products by, xxii, xxvii, 4–5, 130, 144–45; winter living conditions and activities of, 18–19, 51, 170n90; work for Qallunaat whalers by, xviii–xix, 27, 111, 114–17, 125–31, 136–37. *See also* Cumberland Sound

Inuit (as term), xi–xii

Inuit Circumpolar Council, xii

Inuit Heritage Trust, 20, 160, 166n11

Inuit in Cumberland Sound, 2–3; caribou hunting of, 131–35, 191nn71–72; Christianity among, xxvi, 16, 39, 115, 120–25, 175n42, 189n23, 189n25; dispersed communities after commercial whaling of, 136, 144, 150, 193n92; early encounters with Qallunaat of, 110–11; floe-edge whaling of, 125–31, 190n40, 190nn42–43; food shortages and hunger among, 141–45; migration to Cumberland Sound of, xvi, xxi–xxii, 111; Pangnirtung settlement of, xix, 149–50, 158–60; population increases among, 144; Qallunaat relatives of, 137–41; RCMP jobs of, 139–40; recent legal whale hunting of, 145–48, 194nn117–18; shamanic practices of, 122, 124–25, 146–47; shipwrecked materials and, 138; trade with Qallunaat of, 2, 111, 116; women's hunting activities in, 128, 131, 190n53. *See also* Qallunaat in Cumberland Sound; seasons in Cumberland Sound; whaling industry

Inuit in the United States, 35–76, 154; adaptation to American life of, 46–48, 67–69, 72, 75–76; American stereotypes of, 42–50, 174n18; authoritarian attitudes towards, 36, 59–65, 68, 178n116; conceptions of home of, 35–36, 41, 55–59, 177n87; descriptions of America and other Qallunaat lands by, 39, 49, 68, 76, 154; diseases and deaths of, 36, 50–54, 70–74, 180n165, 180n170; economic challenges of, 36, 65–69; itinerant lives of, 36, 50–59; language use by, 44, 70; numbers of, 36–38; returns to the Arctic by, xiv, 55–56, 59–60, 65, 69, 75–76, 154; scrutiny and judgment of, 36, 40–50. *See also* Ebierbing, Hannah and Ipiirvik

Inuit Knowledge Working Groups for Auyuittuq National Park, 157

Inuit Qaujimajatuqangit Principles, 151, 195n4

Inuit Tapiriit Kanatami (ITK), 150, 164n42

Inuk (as term), xxvi

Inuktitut language, 44, 150; comparison to Inuktut of, 162n12; Qallunaat stereotypes of, 44; terms for home in, xx–xxi, 163n30; terms for non-Indigenous outsiders in, xv; terms for seasons in, 2–3, 90; translation of, 158; use of place names in, xxi, 20, 112, 126, 160

Inuluapik, xvi–xvii, 47, 48, 153

inummariit, 44–45, 174n33

Ipellie, Alootook, 45

Ipiirvik. *See* Ebierbing, Hannah and Ipiirvik

Isabella, 13, 15

Ishulutaq, Elisapee, 21f, 126, 129–30

Isigaittuq, 65, 66f, 70

Ittinuaq, Ollie, 51

Ittuluk, 75–76

Ivalu, xxvi

Jewell, Winfield, 92

Joamie, Aalasi, 70

Joamie, Eric, 193n92

Johnnibo, 68–69

Kalaallit, xi–xii*n*; as expedition employees, 78, 80, 86, 100–103, 107, 184n72; Qallunaat ancestors of, 93, 183n49; responses to environment by, 88–94

Kanajuq, 152–54, 195n8

Kane, Elisha Kent, 1, 46, 83, 99

Kangersuatsiaq (Prøven), 88–91, 183n49

Kappianaq, George Agiaq, 158

Kaunaq, 16

Keenainak, 126f

Kekerten. *See* Qikiqtat

INDEX

Kilabuk, 152–54, 195n8
Kilabuk, Josephee, 127f
Kinnes, Robert, 136
Kislingbury, Frederick, 77–78, 79f, 87, 99, 184n72
Klinkenborg, Verlyn, xx
Kowjakuluk, 30
Kudlago, xiv, xvi, 46–47
Kupaaq, Michel, xx
Kuptana, Rosemarie, 61

Lady Franklin Bay expedition, xix, 78–109; *Arctic Moon* newspaper of, 88; Christmas celebration of, 88, 92; deaths of members of, 78, 98–99, 103–6, 185n104; diet of, 87, 100–101, 103–6, 186n128, 187n148; expedition cabin of, 80, 81f, 84f, 85, 93, 95; failed resupply and rescue missions to, 84–85, 95, 107; first winter solstice of, 88–94; impact of sunlight on, 99–103, 185n112; impact of winter darkness on, 81–87, 99, 182n31; Kalaallit employees of, 78, 80, 89–90, 100–103, 107, 183nn53–54, 184nn72–73, 186n139; Kalaallit responses to environment of, 88–94; military culture of, 80; notions of home among members of, 78–80; Qallunaat members of, 79f; Qallunaat responses to environment of, 85–87; rescue of survivors of, 80, 103–4; retreat to Cape Sabine of, 78, 95–99; scientific goals of, 78; sensationalizing of, 104, 106–9
Lady Greely, 95f
Larcom, Lucy, xxii
late fall (*ukiaq*), 2, 10–16, 141–45, 167n28
lice, 72
Life among the Qallunaat (Freeman), 162n13
Lincoln, Abraham, 52
Living Mountain, The (Shepherd), 194n125
Lockwood, James Booth, 83–84, 93, 106
Long, Francis, 100–103, 105, 107–8, 185n121, 187n178
Luciano, Dana, 54–55

Macfarlane, Robert, 147, 194n125
maps of Cumberland Sound, xviif, xviiif
Martin, Keavy, 162n13
mattak, 144–45
McClintock, Francis, xxvii
McGuinn, John, 140–41
McLellan, 3–6, 13, 25–28, 33, 38, 166n16
"memoryscape," 20, 169n80
Merqusaq, xxvii–xxviii
Mike, Jamesie, 97–98, 131, 138, 152–53, 192n85, 195n9
Mike, Louee, 56, 177n87
Mike, Meeka, 166n8
Milne, W. F., 189n13
Milwood, 17, 21, 30, 173n6
motorized boats, 129, 191n59
Mutch, James, 193n110

Nakashuk, Margaret, 110
Nakashuk, Saullu, 107, 144, 192n91
Nashalik, Inuusiq, xii–xiii, xv, 129–31, 134, 169n80
National Inuit Suicide Prevention Strategy (Inuit Tapiriit Kanatami), 150
Nettilling Lake, 131–34
New Boston Aquarial and Zoological Gardens, 45–46, 174n36, 179n136
non-Indigenous outsiders (Qallunaat). *See* Qallunaat (non-Indigenous outsiders)
Northwest Passage, 1
Nowyook, Nicodemus, 116, 127
Nuijaut, Evie, 120–21, 125, 189n22
Nunavut, xvi, 151, 157–59; Inuktut language of, 162n12; place names in, 20, 112, 126, 160; whale hunting in, 145–48, 194nn117–18. *See also* Arctic; Cumberland Sound
Nunavut Arctic College, 164n42
Nutt, Commander, 41
Nuttall, Mark, 20, 169n80

Onisimus edwardsi, 104–5
oral history research, 157–60
Ottokie, Uqsuralik, 118
overwintering whalers, xviii, 2–6, 10–27; activities and entertainment of, 20–22; clothing needs of, 14, 168n50; death rates among, 17, 168n64, 168n66; descriptions of the Arctic by, 17–20, 32, 168n63; diet and living conditions of, 10–16, 22–25, 168n46, 169n102; freeze up of boats of, 10–11, 167n28; homesickness among, 114–17, 189n13; illness and injury among, 17, 22–26, 31, 168n64, 168n66, 169n102; learning of Inuit language and traditions by, 21–22; social interactions with Inuit of, 14–16, 116–17

Pangnirtung, xix, 130f, 137f, 139–40, 149–50, 158–60
Papatsie, July, 166n8
Parry, William, 182n31
Patdloq, xxviii
Pavy, Octave, 89f
Peck, Edmund, 121
Penny, Margaret, 168n63
Penny, William, xvi–xvii, 188n8
Pitseolak, Koodloo, 124, 190n53
Pitseolak, Markosie, 115, 127, 190n40
Pitsiulak, Lipa, 146f
Piugaattuk, 145
place, xxiv–xxv, 112. *See also* home
place names: in Cumberland Sound, xvi, xviiif, xxi, 20, 110, 112, 126, 157, 160, 163n19, 166n11; in Nunavut, 20, 112, 126, 160
Polaris expedition, xviif, xxvi–xxviii, 75; ice floe experience of, xi–xiv, 33, 74, 161n2, 161n5; Inuit participants in, xi–xiv, 69, 183n54; as popular narrative, xiv, 74

pulmonary diseases, 70–74
"Punny" (*panik*/Isigaittuq), 65, 66f, 70

Qalasiq, Salome Ka&&ak, xxi
Qallunaat (as term), xv, 162n13
Qallunaat (non-Indigenous outsiders): conceptions of home of, xix–xxiii, 32–34, 164n38, 164nn44–45, 172n144; environmental history of the Arctic and, xxiv–xxvi, 164nn44–46; racism of, 60, 63, 67, 92–93, 159; stereotypes of Inuit of, xxvi, 14–16, 42–59, 92–94, 142–43, 168n46, 174n18; stereotypes of the Arctic of, xiv–xv, xx, xxvi, 1–2, 8, 10, 17–20, 32, 78–80, 108–9, 151, 168n63, 169n72, 182n31; time concepts of, xxv–xxvi, 39; Victorian notions of womanhood among, 15–16. *See also* Canadian government policies; colonialism; United States; whaling industry
Qallunaat in Cumberland Sound, 1–34; early encounters of Inuit with, 110–11; winter experiences of, 2–6, 10–27, 114–17, 189n13. *See also* Inuit in Cumberland Sound; overwintering whalers; seasons in Cumberland Sound; whaling industry
qalupaliit, 146–48
Qalupaliq (Pitsiulak), 146f
qammat dwellings, xvi, 4, 58, 117–20
Qaqqaqtunaaq, 139f
Qikiqtani Inuit Association and Qikiqtani Truth Commission, xxiv, 149, 164n42, 193n92
Qikiqtat: Christianity at, 121–23; Inuit living at, 113–19; present day use of, 145; as whaling and trading station, 27, 126–28, 135. *See also* Cumberland Sound
Qikittat (Kilabuk), 127f
Qillarsuaq, xxvi–xxviii, 97, 165nn53–54
Qimiqsuut, 3, 166n11
Qitsualik-Tinsley, Rachel Attituq, xii, xxi, 61
qivittut/qivittoq, 90–92, 183nn60–61
Quayle, William, 3

Rak, Julie, 162n13
Ralston, David, 96, 101
research methodology, 157–60
resistance, 150–51
Rice, George, 93, 97, 99, 101–2, 105, 184n73
Riffenburgh, Beau, 104
ringed seals. *See* seals (mostly ringed seals)
rock tripe (*Umbilicaria*), 105
Ross, W. Gillies, 24, 171n116

Schneider, Private, 88, 92, 94
Schwatka, Frederick, 62
Scott, Robert Falcon, 24
scurvy, 22–26, 83, 87, 169n102
Seager, R. W., 51, 64

seals (mostly ringed seals): as food, xvi, 4, 6, 21–26, 100–102, 143–44, 153; hunting of, 22, 61, 67, 89f, 113, 119f, 120, 125, 129f, 133, 137, 142f, 158, 190n40, 190n53; protests against hunting of, 150; uses for skins of, 4–5, 10, 21, 30, 33, 58, 69, 104–5, 116–17, 121, 128, 130; uses of oil from, 18, 71, 73, 88, 138
seasons in Cumberland Sound, 2–31, 166n8; early fall (*ukiaksaaq*), 2, 7–10, 135–41; early spring (*upingaksaaq*), 3, 22–27, 120–25; late fall (*ukiaq*), 2, 10–16, 141–45, 167n28; spring (*upingaaq*), 2–3, 27–31, 125–31, 190n40; summer (*aujaq*), 2, 3–7, 131–35; winter (*ukiuq*), 3, 16–22, 113–19
Second Great Awakening, xxv–xxvi
Sedna, 122, 189n28, 189n30
settlers. *See* Qallunaat (non-Indigenous outsiders)
Shepherd, Nan, 194n125
Simpson, Audra, xv
skin boats, 132, 191n71
snowmobiles, xix, 134, 153, 195n10
sperm whales, 29
Spicer, John Orrin, 33, 75
spring (*upingaaq*), 2–3, 27–31, 100, 125–31, 190n40
starvation. *See* hunger
Sterry, William, 6, 25–26, 153, 171n115
Storm Beaten drama, 108
Stuhl, Andrew, xxiv
Sullivan, John, 8–10, 14, 22–23
Sulugaalik, 124
summer (*aujaq*), 2–7, 131–35
summer solstice, 103
Sundays. *See* time
sunlight, 99–103, 185n112
supernatural, xxvii–xxviii, 55; *qalupaliit*, 146–48; *qivittut/qivittoq*, 90–92, 183nn60–61; Sedna, 122, 189n28, 189n30; *tuurngait/tuurngaq*, 20, 114, 121–25

Tagaq, Tanya, 150
Tanner, Henry, 185n104
Tarralikitaq, 43f, 50–57, 70, 73, 176n75
tattoos, xxvi
Taylor, John, 135–36, 138–40, 158, 192n85, 192n87
terre et les rêveries de la volonté, La (Bachelard), 87
Tessuin, 27
time, 110–13, 164n46; Inuit conceptions of, xxv–xxvi, 18, 39, 110, 115, 125, 146; in Inuit seasonal distinctions, 2–3, 90, 113, 166n8; in Inuit storytelling practices, 111–12; Qallunaat conceptions of, 39
tobacco, 110–11
Tookoolito. *See* Ebierbing, Hannah and Ipiirvik
Trott, Christopher, xxi, 121, 123, 175n42, 189n31
Truelove, 6
Tuan, Yi-Fu, xx

INDEX

tuberculosis, 70–74
tupiit (sealskin tents), 58, 117, 128
tuurngait/tuurngaq, 20, 114, 121–25
Tyson, George, xi, 6, 13–15, 181n4; alleviation of scurvy by, 25; attachment to Cumberland Sound of, 32–33; on early spring in Cumberland Sound, 22; on overwintering, 4, 17, 20, 33, 166n16; on whale hunting, 27–28, 30

Ulloriaq, Inuuterssuaq, xxviii
Unirsagaaq, 125–27, 190n47
United States: conceptions of time in, 39; economic relationships in, 66–69; Inuit living in, 35–76; mourning practices in, 54–55; the nuclear family in, 46, 175n42; pulmonary disease in, 70–71; racial hierarchies of, 63, 159; social and power inequalities in, 59–65; train travel in, 52, 80; views on health in, 71–74. *See also* Inuit in the United States; whaling industry
Uqi, xxvi–xxvii
Uugaq, 68–69, 154
Uummannarjuaq: Christianity at, 120–21, 123–24; Inuit living at, 131, 137–38, 142*f*, 143, 190n53; as whaling and trading station, 12, 18, 27, 114, 127, 129*f*, 135, 188n8. *See also* Cumberland Sound

Veevee, Pauloosie, 125–26, 137, 144–45, 190n40
Veevee, Rosee, 119
vitamin C, 22–25, 171n113

Watson, Samuel, 23
weeks. *See* time

whaleboats, 8*f*, 9
whaling industry, xvi–xx, 1–34, 163n22; beluga whales in, 31, 40, 128–31; as colonial enterprise, 130–31, 151; commodities produced by, xix, 5, 28–31, 166n16; compensation of crew in, 3–4, 6, 15, 27, 167n19, 171n119; crew demographics in, 4, 166n6; in Cumberland Sound, xvi–xx, 31–34, 110–11, 114–15, 127–29, 163n22; desertion in, 7–10; ending of, xix–xx, 31, 127–29, 136–37, 144–45; hunting practices of, 5, 27–31, 166n16; Inuit upheaval by, xxi–xxii, 113; Inuit workforce of, xviii–xix, 27, 111, 114–17, 125–31, 137–38; overwintering practices of, xviii, 2–6, 10–27, 114–17, 168n46, 168n50, 168nn63–64, 168n66; profit motive in, 5–6, 25, 30–31, 115, 166n16; ships as home in, 32–34, 172n144; shipwrecks in, 11–12; trade and interactions with Inuit in, 2, 14–15, 32–34, 111–17, 137–38; vessels used in, 172n132. *See also* overwintering whalers; *and also names of particular ships, e.g.,* Antelope
winter (*ukiuq*), 3, 16–22, 113–19. *See also* overwintering whalers
winter solstice, 88–94
Wiser, George L., 23
Womack, Craig, 153
women: hunting activities of, 128, 131, 190n53; Inuit women partnering with whalers, 15–16, 34, 115, 138–39; as shipboard tourists, 172n144; Victorian notions of, 15, 88; violence involving, 16
World's Columbian Exposition of 1893, 108